Fish Stock Assessment

FAO/WILEY SERIES
ON FOOD AND AGRICULTURE

Volume 1: **Fish Stock Assessment: A Manual of Basic Methods**
J. A. Gulland

Fish Stock Assessment

A Manual of Basic Methods

J. A. GULLAND

Marine Resources Service
Fishery Resources and Environment Division
Food and Agriculture Organization of the United Nations
Rome, Italy

A Wiley–Interscience Publication

JOHN WILEY & SONS

Chichester · New York · Brisbane · Toronto · Singapore

Reprinted August 1985

Library of Congress Cataloging in Publication Data:

Gulland, J. A.
Fish stock assessment.
(FAO/Wiley series on food and agriculture; v. 1)
'A Wiley–Interscience publication.'
Includes bibliographical references and index.
1. Fisheries—Statistical methods. 2. Fish
populations–Measurement. I. Title. II. Series.
SH331.5.S74G84 1983 333.95′611′0287 82-13665

ISBN 0 471 90027 3

British Library Cataloguing in Publication Data:

Gulland, J. A.
Fish stock assessment.—(FAO/Wiley series on
food and agriculture; v. 1)
1. Fishery management—Mathematical models
I. Title II. Series
333.95′617 SH329.M3

ISBN 0 471 90027 3

Photosetting by Thomson Press (India) Limited,
and printed by The Bath Press, Avon

Contents

List of Symbols

Note: Where a symbol is only used infrequently, or in a special content reference is made to the particular section or equation in which it is used.

A	Area
A	Annual death rate (section 4.3.1)
a	Constant, especially in linear regressions
a	Probability that a particular fish will be caught by a particular hook in unit time (equation 2.7)
B	Biomass
B'	Biomass in exploited phase (equation 5.8)
B_∞	Biomass at maximum population (= carrying capacity)
B_p	Biomass of predators (equation 7.11)
b	Constant, especially in linear regressions
b	Selection factor, relating mesh size to length at first capture
b	Degree of density dependence in recruitment (equation 6.11)
C	Catch in numbers
$C_1 \ldots C_c$	Species competing with target species (section 7.4)
c	Ratio of length of first capture to maximum length (equation 5.11)
D, D'	Density of fish on fishing grounds
D	Number of fish dying of disease (section 4.3.1)
$\bar{d}$	Mean depth of a lake (section 7.5)
E	Subscript denoting expected value
E	Probability of ultimate capture, equal to the ratio of fishing to total mortality. Often termed exploitation ratio
F	Fishing mortality; more strictly instantaneous fishing mortality coefficient
$F_1 \ldots F_f$	Species on which the target species feed (section 7.5)
f	Fishing effort
f'	Adjusted fishing effort (equations 2.6, 2.8)
G	Exponential growth rate (section 5.1)
G	Net gain from a change in gear selectivity (section 5.3)
h, h'	Constants in growth equations (section 4.2)
i	Subscript denoting value in a particular year
K	Coefficient in the von Bertalanffy growth equation
K	Selection factor in gill-nets
k, k'	Constants in growth equations (section 4.2)
k	Average fecundity

L	Total life span in the fishery (section 3.5)
L_∞	Maximum length, especially in the von Bertalanffy growth equation
L	Immediate losses incurred in change in gear selectivity, as a proportion of original catch (section 5.3)
l	Length
l_c	Length at first capture
l_d	Length at which a fish leaves the selection range of a gill-net (section 4.4.3)
l_m	Length at which a gill-net operates with maximum efficiency (section 4.4.3)
l_p	Length at partial recruitment, used in calculating effects of discards (section 4.4.5)
l_r	Length at recruitment
l_s	Minimum length or minimum landing size (section 4.4.4)
M	Natural mortality; strictly instantaneous natural mortality coefficient
m	Exponent in general production model (section 3.2)
m	Constant in Richard's growth curve (equation 4.9)
m	mesh size
N	Number of fish
N_R	Number of fish released by a change of selectivity (equation 5.21)
N_k	Number of fish retained when selectivity changes (equation 5.18)
n	Number of hooks in a long-line (equation 2.7)
o	Subscript denoting observed value
O	Number of fish dying of other causes (section 4.2.1)
P	Number of fish dying from predation (section 4.2.1)
p_p	Production of plants (section 7)
P_H	Production of herbivores (section 7.5)
$P_1 \ldots P_p$	Predator species eating the target species (section 7)
p	A proportion, e.g. of hooks, on a long-line occupied by fish (equation 2.7)
p_i	Relative fishing power of the *i*th vessel
Q	Gross long-term increase in catch following change in selectivity (section 5.3)
q	Catchability coefficient
R	Number of recruits
R'	Number of fish at the age of first capture
$R_1 \ldots R_r$	Species affecting the recruitment of target species (section 7.5)
r	Radius of a fishing gear, or of its influence (section 2.3.1)
r	Maximum rate of population growth (section 3)
r	Subscript denoting time intervals
S	Annual survival rate
S	Abundance of spawning stock (section 6.3)
s	Surface area of a fish (section 4.2.3)
t, T	Time

t_c	Age at first capture, mean selection age
t_L	Maximum age in the fishery (section 5.1)
t_0	Constant in von Bertalanffy growth equation
t_p	Age corresponding to partial recruitment length L_p (section 4.4.5)
t_r	Age at recruitment
U	Catch per unit effort (c.p.u.e.)
$U_0 \ldots U_3$	Constants in the expression for yield in weight (equation 5.6)
V	Value of individual fish (equation 4.8)
W, w	Weight
W_∞	Maximum weight, especially in von Bertalanffy growth equation
W_c	Weight at mean selection length
W	Average weight of fish larger than mean selection length (section 5.3)
W_k	Weight of retained catch (section 5.3)
X	'Other loss' rate in analysis of tagging data (equation 4.33)
x	Subscript denoting values in a particular year
Y	Yield in weight
Z	Total mortality, strictly instantaneous total mortality coefficient

Series Preface

The Food and Agriculture Organization of the United Nations (FAO) is one of the world's largest publishers in the fields of food, agriculture, fisheries and forestry. With more than 150 member countries and over 35 years of experience of collecting and disseminating global information, FAO is in a unique position to make available to an international audience statistical data, technical reports and specialized studies—many of them based on FAO's own practical experience as an executing agency for field projects in the Third World—on a wide variety of topics.

Most of these publications are of interest only to highly specialized individuals, however, and it is the aim of this FAO/Wiley series to bring material of more general interest to the attention of a wider audience. We hope that this series will help to spread the knowledge of FAO experts to people throughout the world.

December 1982

Chief Editor
FAO

Preface

The conservation, management, and rational utilization of fish stocks is receiving increasing attention. Part of this arises from the general public concern with the environment, and with fishing as a major use of the marine environment. Part arises from the concern of fishery administrators that most familiar types of fish are becoming fully exploited and the difficulties of managing fish stocks can no longer be avoided by moving to other unexploited stocks. Finally, a major cause of interest arises from the third UN Conference of the Law of the Sea, as a result of which coastal states are gaining control over the fish stocks off their coasts, usually through the establishment of an Exclusive Economic Zone out to a distance of 200 miles. To exploit these resources, and to manage and develop the fisheries, while conserving the fish stocks, it is essential to have accurate information on these stocks, and to assess the effects of fishing and other factors on them. The basic purpose of this book is to provide fisheries scientists with some of the tools to make these assessments.

The assessment of a fish stock must take account of all the relevant factors (changes in the natural environment, changes in competing species, etc.) and not solely the direct effect of fishing on that stock. Nevertheless the analysis of the direct impact of a fishery on a single species is an essential basis for more complex and more realistic analyses.

Thus the main part of the book, after an introductory chapter describing why assessment of fish stocks is needed, deals with the basic single species analyses. Chapter 2 describes the main types of data used. These to a large extent come from the commercial fishery, so that the compilation of data, and subsequent interpretation, raise some different problems from those met in other fields of quantitative ecology. The next chapters describe the methods of analysis, and of estimating the vital parameters (growth, mortality, etc.) used in the analysis. Two basic models are described, the simple surplus-production approaches, treating the population as a single biomass (Chapter 3), and the analytic, age-structured models (Chapter 5). The final chapters discuss the methods used to take account of other factors affecting stocks of individual species and including the interactions between different species.

This manual is directed especially for those working or planning to work in research institutions attached to national fishery departments, who are charged with providing scientific advice for national policies of fishery management and development. At the same time it should be of interest to any scientist interested in the quantitative aspects of fish, marine biology, and quantitative ecology in general.

J. A. GULLAND
1982

CHAPTER 1

Introduction

1.1 THE TASK OF STOCK ASSESSMENT

Fisheries are based on stocks of wild animals, living in their natural environment. These stocks cannot be controlled in the direct and positive way that the farmer or rancher controls his domestic stocks. Nevertheless, the fish stocks are affected by man's activities—and to an increasing extent—and the success of the fisheries depends critically on the state of the fish stocks. All those concerned with making policy decisions about fisheries must take into account, to a greater or lesser extent, the condition of the fish stocks and the effect on these stocks of the actions being contemplated. The obvious example is a government, alarmed by the drop in catches from a particular fishery, which is considering introducing regulations to counteract 'overfishing' and so rebuild the stock and thus allow good catches to be taken once again. Such a government needs to know what effects the different management measures that could be considered would have on the fish stocks and on future catches. Many other organizations need information on the stocks. International banks considering loans to a country to build a fleet of modern vessels need to know what catches the proposed ships will take, and also the effect that these additional vessels will have on the fish stocks, and hence on the catches in the existing fishery.

These decision-makers therefore need scientific advice about the state of the fish stocks. The science of stock assessment is concerned with the provision of this advice. The later sections of this manual deal with the methods of analysing the relevant data in order to prepare advice, but however good the calculations and analyses on which the advice is based, they can be ineffective when it is prepared and presented without some understanding of the problems faced by the decision-makers, and how the scientific advice can help in taking the right decisions to tackle the problems. Practical decisions by governments and others are taken after balancing a range of economic, social, and political factors. Biologists are seldom in the position to say what should be done; their responsibility is to determine the effect of different possible alternative actions, so that others with wider responsibilities can determine which alternative is most satisfactory.

This advisory function does not preclude a high degree of initiative in determining which alternative decisions should be considered, and the range of effects (direct and indirect, immediate and long-term) arising from each alternative which should be analysed. The history of fishery management and conservation, at least from the time of Michael Graham in the 1930s onwards, has been that the scientists, and particularly those concerned with stock assessment, have played some of the leading roles in the consideration of the wider aspects of resource conservation, and in promoting the introduction of those management measures that will best serve long-term social and economic interests. Partly this has been due to the interests of the scientists concerned, but it is also largely due to the nature of the problem. The final implications may be non-biological, but social well-being of the fishermen and the economic success of the fishing industry are based on the fish stock, and often it is the stock assessment scientist, because of his concern with the fish stock (and also to some extent his quantitative approach) who is in the best position to appreciate the likely long-term implications.

For example, if he is dealing with a fish stock suffering from growth overfishing (in the sense define in Chapter 2), the scientist may soon determine that the rate at which the smaller fish in the catch would grow, if they were not caught, exceeds the loss rate from natural causes such as disease and predators. A typical piece of stock assessment work, in the most narrow sense, would then be to calculate that by avoiding catching fish below a certain size and thus allowing them to grow to a better size, the total catch in weight could be increased by some 10 per cent.

This, in itself, is not a very useful result for the decision-makers unless they are also informed of the measures that might eliminate or reduce the catch of small fish and of what other effects, immediate or long-term, might arise from these measures. For example, in a trawl fishery the use of an appropriate size of mesh in the cod-end would allow the small fish to escape, but the enforcement of this mesh size would cause difficulties if the same fishermen wished to fish for another, and smaller species. Further, the measure, if successful in its immediate objective, and if no other measures are taken, is likely to cause additional effort to enter the fishery, attracted by the rise in catch rate. This additional fishery will tend to reduce the stock, thus bringing catch rates, and even total catch, back towards their original level before the measure was introduced. All these are matters for the stock assessment scientist to think and write about. Though this manual is intended to be concerned mainly with the calculations needed to estimate that the release of N fish of a certain size will, in the long run, increase the total catch by x per cent and this can be done by increasing the mesh size to y mm, it is also intended to show how these estimations fit into the larger framework of improved resource use.

This framework will develop, and the nature of relevant aspects of fishery science and stock assessment will expand and change, as the fisheries themselves develop. Following Regier (1976) it is possible to draw up a tabulation distinguishing stages in the development of the use of the fish resource in a given body of water, or ecosystem (see Table 1.1). At each stage Table 1.1 shows the fishing activities being carried on and the general biology research relevant to these activities, and within that research the more specific stock assessment topics. Each stage represents a widening of the scope of the activity shown in the column concerned (fishery, general research, or stock assessment). Therefore,

Table 1.1

Resource use	Catches	Relevant biological studies	Stock assessment activities
1. Exploration, trial fishing	Low	General description of main stocks (taxonomy and distribution)	Order of magnitude estimate of main stocks
2. Developing fisheries on most profitable stocks (MPS)	Moderate and increasing from MPS	More detailed description of life history of MPS	First assessment of potential of MPS
3. Intensify fishing on MPS, and start fishing on less profitable stocks (LPS)	Moderate to high, particularly from MPS	Population dynamics of MPS. Identification of main inter actions between stocks	Detailed establishment of yield curves for MPS and of measures needed to attain desired point (MSY,[a] OSP,[b] etc.) on these curves
4. Intense fishery on all marketable stocks	High, with possible decreases from vulnerable stocks	Population dynamics of all stocks. Study of interactions, and of structure of complete ecosystem	Yield curves for all stocks, and estimates of main interactions
5. Complete resource management (possibly following period of 'overfishing')	High	Ecosystem studies and dynamics	Assessment of the effects of action in any stock/fishery on any other stock/fishery

[a] Maximum sustainable yields.
[b] Optimum sustainable population.

the stock assessment studies of one stage are not sufficient to deal with the problems of a later stage—the detailed species-by-species analysis typical of the stock assessment work appropriate to a fishery in stage 3 is insufficient to deal with the fisheries in stage 4, where each species is heavily fished, and the interactions between fisheries on different species may be very significant. Since it takes time—and often considerable time—to collect and analyse the data appropriate to a given level of stock assessment work, this means that at any given moment scientists should be preparing to tackle the assessment work appropriate to the stage *following* the one in which the fishery is currently placed. This is unfortunately opposite to current practice, which too often deals most effectively with the problems of earlier stages—generals are not the only ones who learn best how to win the battles of the war before last.

Since fisheries in most parts of the world have arrived at, or are moving towards, the later stages of Table 1.1, where most of the resources in a given region are moderately to fully exploited, stock assessment work must also be mainly concerned with these later stages, and be considered with the operation of the whole ecosystem and of the interaction between different resources and different fisheries. The importance of these whole-system studies, outlined in later sections of this manual does not, however, reduce the importance of the single-species studies described in the earlier sections; the more complex studies need these as vital elements in the complete analysis. For example, until a good analysis of the effect of fishing for cod on the cod stocks of the Newfoundland Banks under conditions of a constant capelin population is available, it is difficult to evaluate what the effect of fishery for capelin (one of the principal foods of cod off Newfoundland) would be on cod stocks and catches.

1.2 MODELS

In studying the state of the fish stocks and the effect of fishing on them, the fishery biologist should carry out his analysis in quantitative terms. To do this he must use mathematics, and to use mathematics the complexities of the real situation must be replaced by more or less simplified and abstract mathematical models. Such models may be used to represent both quantities of interest (abundance of the population, size of the individual fish), and the relation between these quantities.

In their simplest form, mathematical models are regularly used by biologists; for instance, it is commonplace to represent the size of a fish by the number of centimetres between the tip of its jaw and the end of its caudal fin. This model conceals many factors about the actual fish—whether it is fat or thin, or whether it is a cod or a tuna—but enables many analyses to be carried out—e.g. the construction of a length-frequency distribution of a sample of the fish population.

The value of a model may be judged by its simplicity and the closeness with which events or values predicted by the model fit the actual observation. A model cannot be considered as being either right or wrong, but as giving a satisfactory fit to the facts over a wide or narrow range of situations. A good model is one that is simple, but gives a good fit over a wide range.

The best test of a model is its usefulness in prediction; in this sense prediction covers not only the prediction of future events but also any values or events not considered in proposing the model. Thus, a model describing the growth of cod in length may be proposed from the analysis of data of mean length at age; it will be a more useful model if, subject to the fitting of a minimum number of constants, it can be used to predict (estimate) the mean length at age of haddock, or any other species.

All the models described in this manual have proved to a greater or lesser extent to be useful models in that they have provided useful quantitative descriptions of events in various fisheries. Most have also been successful in making predictions, in the sense used above. A good example of the testing of a model by a successful prediction is the analysis of the Antarctic whale stocks. At the 1963 meeting of the International Whaling Commission there was considerable discussion on the quota, in terms of numbers of whales caught, to be set for the 1963/64 season. The Commission's Committee of Three Scientists had devised a model for the population of Antarctic fin whales, taking into account the probable rates of mortality and recruitment, etc. On the basis of this model the Committee had recommended that in order to rebuild the depleted stocks, not more than 5000 fin whales should be killed. It also noted that if this recommendation was not followed, and whaling activities were continued at the 1962/63 rate about 14 000 fin whales would be caught, equivalent, including catches of other whales, to some 8500 Blue Whale Units (BWU). Members of the Commission were unwilling to make so large a reduction in the quota from that for the 1962/63 season (15 000 BWU) and therefore agreed on a quota of 10 000 BWU for 1963/64. This gave no effective restriction on whaling activities, and the catches were very close to the predicted values (13 870 fin whales, and a total of 8429 BWU).

Such close agreement between observed and predicted values was partly a matter of luck, that is, certain assumptions made in constructing the model were quite closely fulfilled. In particular, the model assumed that some factors influencing the catch, such as the weather and the skill of the gunners were, in the 1963/64 season, close to the average of previous seasons. These assumptions were nearly fulfilled, but a slight difference in weather or in the skill of a few gunners could have increased the difference between observed and predicted catches to perhaps 300 or 400 animals. The closeness of the agreement did not prove that the model is correct, or that a slightly different model might not give a rather closer fit, but does prove that the model used can predict within useful limits the results of one pattern of whaling activity. Presumably the model could also predict the results of other patterns of activity, and in particular the result of a severe restriction in catches for a period long enough to allow the stocks to rebuild. Therefore the model serves as a usable basis for managing the whale resources.

As in the whale example, most models include simplifying assumptions about the factors which are not of immediate interest (e.g. weather) usually stated, when stated explicitly, in the form 'assuming the conditions are constant'. This should not be taken to mean that the validity of the model depends on the constancy of these conditions, and that the model should not be used if conditions vary. Rather it means that for the purposes for which the model has been constructed, the variations caused by these conditions are not of primary interest, and will be ignored.

This manual is mainly concerned with studying the effect of fishing on the stocks and on the catches and particularly with those aspects of this study which are important in advising those taking decisions on fishery policy (making investment, setting regulations, etc.). Most of these policy-makers require a long-term view of matters, and this manual is, therefore, essentially concerned with the long-term effects. Also, when predicting the catch from a given pattern of fishing the important question is often not the absolute amount of catch but the catch relative to that which would have been taken with some other pattern of fishing. Many fluctuations and variations are therefore irrelevant to the main purposes of this manual. For instance, very often the average catch per trawl haul is taken as a measure of the abundance of the stock; the actual catch taken in one haul will depend on a large number of factors, for example the size of the trawler, the skill of the skipper, the precise ground, the season, the weather, and the time of day, some of which may have a considerably greater influence on the catch of that particular haul than the overall abundance of the stock. Fish population dynamics is, however, seldom concerned with studying or predicting the catch of one haul, but is more concerned with the average catch over a period, say a year. Most of the factors affecting the success of a single haul will have average annual values which are effectively constant from year to year, and thus can be ignored; only if these annual values vary, and particularly if they show a trend (e.g. an increase in the average size of trawler) may they have to be studied in detail.

When comparing the catches from two different patterns of fishing, for example trawling with large or small meshes in the cod-end, even some year-to-year fluctuations can often be ignored. For instance, variations in recruitment may make big differences to the catch in any one year, but will not in general affect the conclusion that, say, with the present amount of fishing the use of meshes of 120 mm in the cod-end will give sustained catches 5 per cent larger than those taken when meshes of 110 mm are used.

Any model will eventually have to be replaced or modified, possibly merely by adding a little complexity to take into account further factors (e.g. weather conditions in the whale example above) to attain rather greater precision, or perhaps being replaced by a different model. In any case, the process of construction of a model, testing it, and modifying or replacing it is an essential part of the study and eventual understanding of the dynamics of fish populations, or indeed any subject of scientific study. The process of model construction is the complement of the collection of data, and in fact it is only by constructing and using models that it is possible to decide what data should be collected.

A particular model is useful at a given time in the study of a fishery to the extent that it can be used with the information available, and can help with the current scientific problems. The models used are therefore likely to change as the information and problems change. Also, the uses of a particular model are likely to change. Typically, a model goes through the common stages of growth and development, maturity, and old age. In its early stages a model (say, of the dynamics of a population of fish of a given species) can serve only as a qualitative description of what is happening; it cannot be used to make quantitative predictions, for example of the effect on the catch of increasing the amount of fishing by 10 per cent, but does provide a conceptual framework on which refinements of the model can be made, and the collection of relevant information planned. With these refinements and further supply of information the models can be made more realistic, and can be used for predictions, quantitative analysis, and the provision of specific advice to decision-makers. The models may then be considered as mature. As they require more detailed advice and more information becomes available, further modifications to the models will be desirable, and up to a point will be possible. At some point further modifications will be impracticable, and a new model, or family of models, will have to be introduced.

A number of models are touched upon, in various detail, in this manual. Between them they represent most of the stages mentioned above. The simplest models are the production models discussed in Chapter 3, which treat a fish population as a single entity, subject to simple rules of increase and decrease. These allow analyses to be made when only very little information, mainly on catch, abundance, and amount of fishing, is available, but cannot readily be adapted to take account of detailed biological information on the fish, or to provide advice on the effects of detailed adjustments to the pattern of fishing, for example on changes in the size of fish caught. To this extent production models are out of date for those fisheries for which detailed information is available or for which detailed advice on the pattern of fishing is required. For these fisheries more complex models are needed. However, as fast as the problems of existing fisheries increased in complexity to the point where production models are insufficient, new fisheries develop with urgent problems but a poor supply of information. For these, production models are often the most suitable, and these models are certain to have important uses for many years to come. The family of models described in Chapters 4–6 give consideration to the separate characteristics—growth of the individual fish, mortality due to fishing and natural causes (disease, predation, etc.), and reproduction—which determine the increase and decrease in abundance of a fish stock. These models, in the forms which allow the parameters to vary in accordance with the density of the stock (see Chapter 6) are the most complex and realistic of those models which are sufficiently developed to provide predictive and quantitative advice. However, as fisheries exploit a wider range of species, and the general impact of man's activities (including various forms of pollution) on the aquatic ecosystem becomes more pervasive, even these models, which treat each species more or less in isolation and are mainly restricted to looking at steady-state situations, are becoming less satisfactory. Models are needed that can examine the ecosystem as a whole or can deal with non-steady-state events. While these exist, they are still little developed (see Chapters 7 and 8), and as yet cannot in general be used to produce predictive or quantitative analyses. However, they are already useful in providing some qualitative impression of the results of different actions, and in suggesting which lines of investigations would be most productive and which data would be most valuable. For the present, most quantitative stock assessment work is still based on the single-species models, often progressing from production to analytic models as data on the fishery accumulate (with some modification of the latter to take account of some other directly interacting species—competitors, food, or predators). In the long term, stock assessment studies must become increasingly concerned with ecosystem models. Beyer (1977) gives a succint statement of the principles of the different models. The reader must therefore be warned that this manual does not attempt to provide a comprehensive guide to everything involved in stock assessment. That part of the picture dealing with multi-species and ecosystem models is still too unclear and changing too fast to be carefully described in a manual of this type.

1.3. THE EFFECTS OF FISHING ON A POPULATION

The effects of fishing on a stock of fish of a particular species and the resulting catches from that stock are, all other things being equal, very simple. When fishing starts the stock is abundant, and the catches taken by the individual fisherman or by the individual vessels are reasonably high, although because there are few vessels and fishermen the total catch is low. Often in the very early period catches are particularly low because the fishermen are inexperienced or the vessels inefficient. Therefore, there is often a period in the first few years of a fishery when the catches per vessel tend to increase because efficiency increases and the fishermen learn the best grounds to fish, and the best way to adjust their gear to the peculiarities of local stocks and local conditions. This 'learning' effect can often obscure or reverse for a period the decrease in catch rates more typical of a developing fishery. The effect is likely to be particularly noticeable and cause particular difficulties in analysis (because data from the critical early years in the development of the fishery cannot be readily compared with data from later years) in the case of fleets of advanced mobile vessels moving into a new area or on to a new stock. In these fisheries it may be useful to apply a learning factor to adjust for the lower efficiency in the first few years of the fishery. Factors used have been as high as 50 per cent per year in each of the first two years (Brown *et al.*, 1976).

Apart from possible effects of learning, in the early stages of a fishery the catches are usually small and will have little effect on the stock. The abundance of fish, and the average catches of the individual fisherman or vessel will, other things being equal, remain about the same. This means that as the fishery develops, and the number of fishermen increase, the total catches will increase about in proportion to the amount of fishing (as measured, say, by number of fishermen) (see Fig. 1.1, lower left-hand part of the curve, section OA).

This happy state of affairs cannot go on for ever. No stock of fish is unlimited—though governments, industries and fishermen often behave during the growth phase of a fishery as though they were —and as the amount of fishing increases it will begin to affect the stock, reducing its abundance and the average catch taken by a unit amount of fishing (e.g. an individual fishermen or fishing vessel). As a result an increased amount of fishing will give less than a proportional increase in catch, and the curve of catch, as a function of the amount of fishing, will flatten out (Figure 1.1, section AB). Ultimately, a point may be reached at which no further increase in catch can be achieved, and more fishing will actually decrease the catch (Figure 1.1, section BCD).

At its simplest, stock assessment is no more than determining for the stock of interest the nature of the curve as in Figure 1.1, relating the catch to the

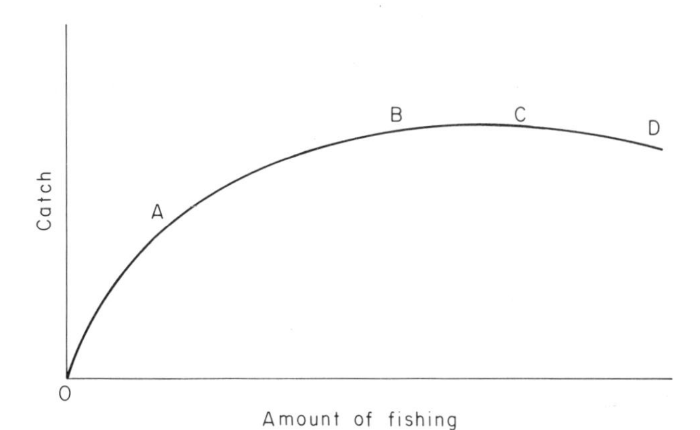

Figure 1.1 Generalized equilibrium relation between the amount of fishing and the catch from a stock of fish

amount of fishing, and the present position of the fishery on the curve. If a fishery scientist is fortunate enough to have information on the catch and the amount of fishing over a period covering a wide range of stock sizes, and to be dealing with a stock that is well behaved (in the sense of lacking the complications discussed below), then the work of assessment need be no more complicated than plotting the catch against the amount of fishing (see exercise 1.1). Chapter 3 discusses more sophisticated methods of analysis using essentially the same approach—observing what the relation is between the amount of fishing and the catch, either purely empirically or making some assumption of the form of the relation, for example that it will be a simple parabola.

Neither this qualitative description, nor the models discussed in Chapter 3 give insight into how fishing affects the stocks or how, beyond a certain point, increasing the amount of fishing can decrease the catch. There are two distinct reasons for this decrease that can, in a given fishery, apply individually or in combination. The first is that the fish are being caught so young that they are not being given a chance to grow to a decent size. Though increased fishing will increase the *number* of fish caught, their average weight will steadily decrease and so, ultimately, will the total weight. This may be referred to as growth overfishing (Cushing, 1972). The mathematics of the effect are illustrated in Table 1.2 (see also exercise 1.2).

The second and potentially more disastrous reason for falling catches as a result of heavy fishing is that the spawning stock is reduced to too low a level to ensure adequate production of young fish—the recruits to the future fishery. This may be termed 'recruitment overfishing'. Because of the enormous fecundity of most fish, a handful of females can produce millions of eggs which, if they all hatched and survived, would be more than adequate to ensure future recruitment, recruitment overfishing is often not a problem. Economic or other factors intervene before fishing reached such a high level as to reduce stocks

Table 1.2
Catches under different intensities of fishing, giving total numbers, total weight, and average weight of the individual fish. This shows how, under certain circumstances of growth and survival, increased fishing can produce a smaller total weight

		Catches					
Age	Average weight	Light fishing		Moderate fishing		Heavy fishing	
		No.	Weight	No.	Weight	No.	Weight
1	1	20	20	40	40	65	65
2	3	15	45	25	75	25	75
3	10	10	100	15	150	5	50
Total weight			165		265		190
Total numbers		45		80		95	
Average weight		3.7		3.3		2.0	

to the point at which recruitment is affected. Also, because in many stocks the number of recruits entering the fishery from spawning in a given year (the year-class strength) is highly variable (a difference of two or three orders of magnitude in the case of haddock and some other temperate fishes), recruitment overfishing is not easy to detect when it does occur. That is, it is not easy to tell, at least until a run of poor year-classes has continued for some years, whether a series of years with low recruitment from low spawning stocks is due to a succession of years with poor environmental conditions, or whether it is due to the small spawning stock, or a combination of both.

In summary, therefore, it may be said that growth overfishing is common, but not disastrous, and can be readily detected; recruitment overfishing is less common, but may cause disaster when it does occur, but even after disaster strikes it may be difficult to be sure whether it has been caused by overfishing, or a natural failure in recruitment. In either case the methods of study start in the same way with an analysis of the vital parameters (growth and mortality rates, see Chapter 4); these can be used to calculate the yield to be expected from a recruit-class of a given strength under different fishing patterns, (see Chapter 5). This is sufficient to detect growth overfishing, and to determine (to a first approximation) the kinds of steps needed to rectify matters, but for a closer approximation and to assess the possibility of recruitment overfishing, it is necessary to examine the extent to which the natural parameters, including recruitment/reproduction rate may be related to the abundance of the stock (Chapter 6).

In a simple world in which each fishery operated on a single uniform population of a single species in a uniform environment, unaffected by outside events (including other fisheries), this might be all that is involved in stock assessment. Complications that arise and that need to be tackled at an appropriate stage of any assessment study include:

(i) Few species form single homogeneous populations and most can be divided into several more or less distinct stocks, reacting to fishing more or less independently. The 'Unit stock' being exploited by the fishery under study has therefore to be identified (see section 2.1).

(ii) Few fisheries operate on a single species or stock, but on a mixture (100 or more in the case of some tropical fisheries). The interactions between these species need to be considered.

(iii) Several fisheries may operate in the same region on different species, and the range of species exploited is likely to increase with time (see Table 1.1). Again, the interactions between different fish stocks, and between fisheries need to be considered.

(iv) Factors other than the fishing can affect stocks, both natural (changes in environmental conditions) and man-made effects such as pollution, coastal development. While these can often be considered as random variation around the relations of interest (e.g. between the amount of fishing and the average catch) consideration of these factors can be valuable. At a minimum, an understanding of them can reduce the amount of unexplained random variation, thus making the important relations easier to detect. Often, though, the outside factors are of interest in themselves—for example, a recruitment failure may be caused by a low spawning stock only when there are also below-average conditions for the survival of small fish. Some of these factors are discussed in Chapters 7 and 8.

It may be noted that the phrase 'at an appropriate stage' in the previous paragraph deserves careful attention. The timing of the extension of research efforts and assessment analysis from the simple single-fishery, single-stock uniform environment set of assumptions to the more complex but more realistic assumptions is important. Early attempts at studying the entire system, if attempted at a time when the simple single-species assessment has not yet been made, are likely to produce no more than qualitative and general statements which are unlikely to be helpful for decision-makers. If they are done when resources to carry out research are limited, they risk that insufficient resources are allocated to obtaining quantitative assessments of the single-species sub-systems. Conversely, to continue to concentrate attention on these sub-systems, without considering how they are linked together in the whole ecosystem, risks that the assessments made, though precise, are not realistic descriptions of actual events.

1.4 OBJECTIVES

Stock assessment is concerned with advising decision-makers on the effects of different possible actions. This can only be done well if it is known which effects need to be assessed, i.e. which factors, susceptible to scientific evaluation (total catch, annual variation in catch, average size of fish caught, cost of catching) need to be taken into account in assessing how good or bad the results of a given policy are expected to be. To a large extent the absolute value of some particular output (e.g. total catch) corresponding to a given input (e.g. number of ships) is of less interest than the marginal effects (in the sense used by economists), that is, the *changes* in a given output (e.g. catch) arising from a *change* in a particular input (Gulland, 1968).

The pattern with which the results are distributed in time is also important. Most management measures are concerned with applying controls on the current fishery so that the future fishery will be better, i.e. accepting less output now so that future output will be higher. In determining whether this is a worthwhile exercise both the relative magnitude of immediate and long-term effects, and the time which has to pass before the long-term effects occur, need to be taken into account. A large sacrifice in output may be acceptable if the benefits follow quickly, with only a short delay. Conversely, large long-term benefits may seem hardly worth while if there has to be a wait of many years before the benefits are achieved, even if the magnitude of the short-term sacrifices are quite small.

Another important aspect of the time patterns is the degree of variability in catches, etc. Traditionally, assessments have been expressed as averages—'under average condition of year-class strength and environment, reducing the amount of fishing by 10 per cent will increase the total yield by 2 per cent'—but this is by no means a complete description of the possible effects on yields unless it is accompanied by some indication of the variance about the average value. The economic and social impacts of fisheries taking an average annual catch of 10 000 tons are very different if the annual values remain within a narrow band—say 8000–12 000 tons—than if the average is achieved by the occasional very good year of 30 000 tons (and otherwise only 2000–3000 tons), or by a usual fishery of 12 000–15 000 tons, interspersed by occasional years of complete failure. Usually a steady catch year after year is worth more than a varying catch with the same average value; a year with an extremely low catch needs to be avoided if at all possible, even at the cost of some reduction in the overall average. Since the variability can be, to at least some extent, affected by fishing—more intense fishing tends to increase the variation—a complete assessment of the effect of any action should include some evaluation of the effect on the variability of important quantities.

1.5 MAXIMUM SUSTAINABLE YIELD

A common and traditional objective of management has for a long time been the maximum sustainable yield (MSY), though this has been under criticism for almost as long. The present-day importance of MSY is not so much as an objective to be rigidly followed in reaching decisions, so much as a very convenient concept for use in discussing general management problems. This convenience arises because MSY serves, to a useful first approximation, three distinct functions—a description of the facts of life regarding fish stocks in relation to exploitation, a clearly definable objective of management, and a measure of the success with which a stock is being managed.

Maximum sustainable yield would always be a useful description of how fish stocks behaved if their behaviour could be fully and accurately described by a smooth curve with a clear peak like that of Figure 1.1, and if this curve could always be readily determined. The idea of MSY does emphasize to those not familiar with natural resources (administrators, investors, etc.) the fact that more fishing does not necessarily mean more fish, and that beyond a certain point more fishing can mean less fish. One difficulty in the simple MSY concept is that the actual yield in a particular year can be subject to considerable variation from non-fishery causes. Sometimes this can be clarified by referring to the maximum sustainable average yield (i.e. the yield under average environmental conditions)—MSAY. Another difficulty in attempting to use MSY as a quantitative guide to decision-making is that unless the yield curve of Figure 1.1 has a sharp maximum it can be very hard to determine at which level of fishing the MSY occurs, and thus whether a given amount of fishing is above or below that giving the MSY. This difficulty is particularly great when the yield per recruit (as discussed in Chapter 5) does not change much at moderate to high levels of fishing, and the fall-off in yield at the higher fishing rates, if it occurs, is due to a reduction in recruitment. As discussed in Chapter 6, it can be very difficult to determine, until there has been a drastic decline in recruitment, whether or not changes in adult stock affect recruitment, and if so, to what extent. At the same time it is possible to calculate with good precision and a fair degree of reliability, the relation between yield per recruit and fishing mortality (and hence fishing effort). Using this relation, two other signposts can be established which may be of more practical value than MSY in determining whether the amount of fishing should be increased or decreased. These are the values of fishing mortality, F_{MAX}, corresponding to the maximum yield per recruit, and, $F_{0.1}$, corresponding to the point at which the marginal yield per recruit from an additional unit of effort is 0.1 the marginal yield per recruit at very low levels of fishing, and hence a point beyond which there is little reward in increasing fishing. If recruitment is not affected by fishing, then the amount of fishing giving F_{MAX} will coincide with that giving the MSY. Otherwise F_{MAX} and $F_{0.1}$ differ from the MSY point (F_{MAX} occurring at a higher rate of fishing than MSY and $F_{0.1}$ usually at a lower), but serve as alternative identifiable characteristics of a fish stock which, being more easily calculable with the kind of information commonly available, are more useful in determining action in many practical situations.

As an objective, MSY contains three important ideas—of maximizing some quantity, of ensuring that this can be sustained (by implication more or less indefinitely), and of the physical yield (tons of fish) being an appropriate measure of the well-being of a fishery. These are all reasonable, and many of the criticisms of MSY (especially in relation to management of whales) are actually criticism of how it has been applied, and the failure to give at least as much attention in practice to ensuring that the yield is sustainable, as that it is maximized. The more valid criticisms are that maximizing a quantity is seldom a critical matter (a very high value of yield, or economic return, etc. is usually virtually equally

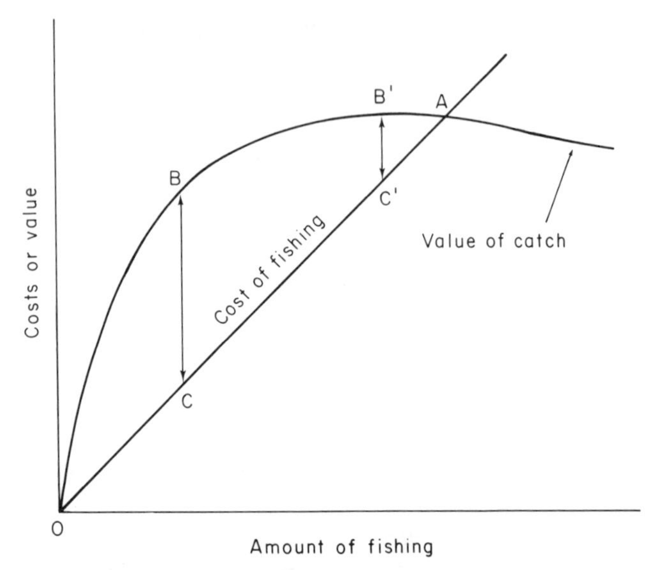

Figure 1.2 Grossly simplified representation of the economics of fishing, to show how the greatest net economic return (value of catch less costs of fishing) occurs at a level of fishing less than that giving the greatest gross catch

satisfactory), and that the magnitude of the physical yield is only one (and not necessarily even the best) of the measures of the well-being of a fishery.

Economic and social considerations are receiving increasing attention. The implications of some of these considerations can be seen by turning the curve of Figure 1.1 into a relation between the costs of fishing and the value of the catch (which will be to a useful first approximation proportional to the amount of fishing and the yield respectively, see Figure 1.2). By adding the line OA relating points of equal costs and value, and considering the height of the curve above this line (which will be equal to the net economic return of value less costs), it can be seen that the economic benefits (BC) at a level of fishing less than that giving the MSY can be considerably larger than those (B′C′) at the MSY point. If economic return were the only measure of success, fishing at the point giving the maximum economic yield (MEY) would be the appropriate objective. Though, as Crutchfield (1975) points out, obtaining a high economic return enables many other objectives (employment, low prices to consumers, high returns to fishermen, etc.) to be successfully pursued, MEY shares many of the disadvantages of MSY in pursuing one factor at the exclusion of others. For this reason the optimum sustainable yield (Roedel, 1975) has been suggested as a concept in management which allows a flexible weighting between various measures of the well-being of a fishery. In this approach the need to define formally a single objective of management (MSY, MEY) which should be applied in all situations regardless of changes in society's needs, or in the understanding of natural resources, has been fairly explicitly rejected.

Finally, the role of MSY as a criterion of management success may be considered. In a simple world it would be sufficient to determine the abundance of a stock relative to the MSY level; if above, the stock is under-exploited and no management is needed; if below, the stock is overfished and action is called for, but if the stock is at MSY then all is well. In practice, holding the population at some determined level may be a necessary, but not a sufficient, condition for management to be successful. Good management needs to take account of many other factors (e.g. from outside the purely biological field, costs of enforcement, balance of benefits and losses between different participants).

It is clear that MSY does not, in the complex situation of most advanced modern fisheries, serve well as an adequate description of the resource, as an objective, nor as an index of management success. Therefore, for these fisheries stock assessment will not often be concerned with the explicit and detailed calculation of MSY, or of the level of fishery or of population abundance at which it might be obtained. Nevertheless, MSY does serve as an approximation to each of these, and as a better simultaneous approximation to each than any other single term. Therefore, it remains as a most convenient concept and bench-mark to be used in general discussions of management, particularly in enabling non-specialists to have some understanding of the problem of how fishing harder produces less fish, and what needs to be done to manage the fishery.

In expressing the results of actual assessments it is preferable to avoid the term 'maximum sustainable yield' because of the many ways it has been used, and the overtones to the term MSY. For example, one of the most common and valuable results of assessment studies in the earlier stages of a fishery is an estimate of how much could be taken annually year after year. This may be determined in a number of ways—from general consideration of the productivity of the area, from trawl surveys, from analysis of catch and effort statistics, etc. It may conveniently be referred to as the potential yield. This brings out the fact that in practice the greatest possible yield may not be taken, because, for example, it is preferred to obtain greater economic benefit by catching slightly less catch at considerably less cost, or because the greatest theoretically achieved yield requires an impracticable combination of a high level of fishing and strictly controlled selectivity of fish of the right size.

1.6 SOME REMARKS ON SPECIAL TOPICS

1.6.1 Fast-growing species

A common concern of those working on fast-growing species, especially in the tropics, with the traditional stock assessment techniques of the type described in this manual and its predecessors, is that they can only be applied to species that live several years. This concern can take several forms. A fishery scientist may fail to use the techniques that could help him, or an administrator (or a scientist) say 'shrimp cannot be overfished because they only live one year'.

There are indeed difficulties in applying the present techniques to many fisheries, especially in the tropics—the large number of species, the lack of easy methods of age-determination—but the relative length of life is not one of the most serious. To a mathematician the point is fairly obvious—the equations are valid whatever period is taken as unit time. Except for differences in the values of the parameters, the same formulae can be used for short-lived or long-lived species. However, if the time-scale is kept at one unit equals one year, some of these parameters become a little hard to comprehend. For instance, the fishing mortality on a shrimp stock may be 8.4, i.e. the annual catch is 8.4 times the average standing stock.

The picture becomes clearer if, for these species, shorter periods, e.g. months or even weeks, are taken as the unit of time. On a monthly basis the fishing mortality becomes 0.7, which is a reasonable figure as an annual rate for a fairly heavily exploited long-lived fish. The factor of 8.4 can then be seen to be no more remarkable than the fact that the catch of plaice in the North Sea over a 12-year period is about 8.4 times the average biomass.

There is, of course, one practical difference between short-living and long-living species—things happen much more quickly. For example, stocks can be expected to recover more rapidly from over-exploitation. The true statement about penaeid shrimp and overfishing is that provided recruitment is not affected by reduction in adult stock, then whatever happens to the shrimp presently in the fishery, and however badly they have been reduced by fishing, all will be well next year when the next brood comes in. In contrast, if anything goes wrong with the stock of cod, it will be 8–10 years after a good year-class is born before the Norwegian fisherman at the Lofoten Islands notice a recovery in their catches.

1.6.2 Use of computers

In a review of the previous version of this manual Paulik (1971) referred to it as an anachronism in the era of computerized ecology and resource management that he saw dawning. Given the advances that have been made in computer technology, and especially the wide availability of cheap computer facilities, the present manual, with its emphasis on techniques that generally involve only simple calculations, risks being even more of an anachronism. This risk has been run consciously. The emphasis in this manual has been on promoting thought, and the understanding of what is involved, rather than on training in the application of rigid formula. The application of computer methods can be too easily used as a substitute for thought, or if thought is used, it is concerned with the programming proceders rather than the biological basis of the program. Nevertheless, once the principles have been understood, it is only sensible to carry out the computations involved, which may be lenghty, in the easiest way, which will normally mean using as advanced computing facilities as are readily available. This has advantages: it will give more time for thought; it will tend to cut down arithmetic errors (though programming errors may occur); and there will be opportunities to try out a greater range of inputs, e.g. looking at yield curves with different values of natural mortality.

What constitutes the most suitable available computing facilities is changing very rapidly, and varies greatly from place to place. While the fishery scientist in a large laboratory in a rich developed country may have direct access, via a terminal in his office, to a large computer with a library of well-established programs of, for example statistical procedures to fit growth curves, the scientist in a small institute in a developing country may have no more than a small hand calculator. While most institutes now have some sort of computing facilities, the gap is likely to remain. No attempt has therefore been made to recommend specific methods or programs, except in a few cases where a particular program (e.g. the ELEFAN program for analysing length frequency data) is an integral part of the method described. However, the Department of Fisheries of FAO is compiling information on the various computing techniques being used to implement the methods described in this volume—and other important techniques related to fish resource assessment. This information, which is regularly updated, can be made available to enquirers. Particular emphasis is being given to techniques using small programmable calculators, and other methods suitable for use in the smaller institutes in developing countries.

1.6.3 Use of length data

The traditional approach to fish stock assessment, as set out in Chapters 3 and 4, is to collect data on the age-composition of the catches (and where possible the stock), estimate the mortality and growth rates, and hence calculate the expected yield under different patterns of fishing. This is done so long as the age of individual fish can be readily determined (e.g. from rings on otoliths). This is not always the case, especially in the tropics, and another approach is needed.

Information on ages of fish is, in itself, not of vital interest, and in many ways size (length or weight) data is of at least equal importance. The original input and final output from the analyses are mostly on terms of numbers and sizes of fish—the age-data being often determined by applying an age–length key to length samples. Age-data only appears in the intermediate stages. It is important because it provides a time-scale against which the processes of growth and mortality can be measured. The length of fish also provides a time-scale. It is less easy to use because it is non-linear, and may be variable (e.g. due to changes in the availability of food). Nevertheless, the direct use of length data, in ways analogous to the uses of age-data does, in principle, afford a method of proceeding when it is not possible to tell the age of individual fish. The methods tend to be laborious, party because of the non-linearity of the time-scale, but computer programs are being developed to deal with at least some of the analyses that are involved. The techniques have not yet been developed to the point at which they can be recommended here as standard practice, with a comprehensive set of procedures to replace analysis of age-data. A number of techniques for using length data in specific circumstances are however described. The more general point to be made here is to emphasize the value of length data in its own right, the importance of regular collection of length data from commercial fisheries, and the desirability of considering various methods of using these data in addition to conversion to age-data.

1.6.4 Doubt and uncertainty in advice

The commonest use of the results of stock assessment is to provide advice to the fishery administrators or the fishing industry when decisions are being taken about the development or management of the fisheries. Especially when management decisions are being considered it is common for the advice to be phrased in specific and exact terms—'The total allowable catch (TAC) should be 11 200 tons' or 'the best mesh size is 85 mm, which will give a long-term gain of 6 per cent in total yield'. The possibility of error is rarely expressed.

There have been good reasons for this. Adoption of management measures in respect of most of the major marine fisheries was, under the old law of the sea, a matter of reaching consensus among a number of countries, often with divergent interests. A presentation of a range of values for the TAC under different hypotheses about, for example, the value of the natural mortality, would almost inevitably mean that the consensus would be reached on the lowest value. A single figure from the scientific advisers has therefore been, under these circumstances, an almost essential prerequisite for reaching effective management decisions.

This is unsatisfactory on two counts. First, most management action involves reaching a balance between different economic and social interests—between a large catch next year, and good catches over a long period into the future, or between high total catches and high catches by the individual fisherman. The fishery biologist is not equipped to make these choices which should be left to the appropriate political body. The new law of the sea, which places most resources under national jurisdiction, makes it easier for the decisions to be taken by the right people.

The second reason, and particularly relevant to the uses of this manual, is that the requirement for a single undisputed figure from the scientists means inevitably that the scientific advice is based on calculations about which there can be little dispute. The following chapters will show that there are very large differences in the reliability with which different effects on fish stocks can be assessed. There may be little argument that increasing the fishing mortality on a given stock by 10 per cent will increase the yield per recruit by about 1 per cent. There is much greater doubt about the extent to which the reduction in stock abundance might affect subsequent recruitment, or the fisheries on other stocks (see Chapters 6 and 7). The result is that too often advice is based on simple yield-per-recruit or similar calculations, and the other, less quantifiable results are ignored. Since these effects are often those that have the greatest long-term impact on fisheries, the scientific advice tends to be rather precise, but wrong, rather than being approximate, but right.

It is therefore highly desirable that when advice is being prepared on the basis of stock assessment studies that such advice is not confined to what appear to be reasonably confident statements, but that the uncertainties, especially those such as the impact on recruitment, that could have a great effect on the fisheries, are fully described. This should in no way diminish the value of the advice. Decision-makers, administrators, industrialists, etc. depend on various sources of advice and types of prediction, for example e.g. on foreign exchange rates or costs of value on future years. Few of those making these predictions would claim high precision—or if they did they would soon be proved wrong—but despite this a reasonable professional estimate of future trends is more valuable than no estimate at all. Biological advice on the effects of different actions should be given and used in the same way.

REFERENCES AND READING LIST

ACMRR Working Party on the scientific advisory function in international fishery management and development bodies (1974) Supplement 1 to the report of the Seventh Session of ACMRR. *FAO Fish. Rep.*, (142), Suppl. 1: 14 pp.

Alverson, D. L. and G. J. Paulik (1973) Objectives and problems of managing aquatic living resources. *J. Fish. Res. Board Can.*, **30** (12), Pt. 2: 1936–1947.

Beyer, J. E. (1977) Basic principles of modelling an exploited marine ecosystem. Contribution to the International Statistical Ecology Programme on Statistical Modelling and Sampling for Ecological Abundance and Diversity with Applications, College Station, Texas, and Berkeley, California, 18 July–13 August 1977.

Chapman, D. G. (1964) Special Committee of Three Scientists. Final report. *Rep. Int. Whal. Comm.*, (14): 39–101.

Crutchfield, J. A. (1975) An economic view of optimum sustainable yield. *Spec. Publ. Am. Fish. Soc.*, (9): 13–9.

Cushing, D, H. (1972) A history of the international fisheries commissions. *Proc. R. Soc. Edinb. (B)*, **73** (36): 361–390.

Graham, M. (1939) The sigmoid curve and the overfishing problem. *Rapp. P.-V. Reun. CIEM*, **110** (2): 15–20.

Graham, M. (1943) *The Fish Gate.* London, Faber.

Gulland, J. A. (1968) The concept of the marginal yield from exploited fish stocks. *J. Cons. CIEM*, **32** (2): 256–261.

Gulland, J. A. (1971) Science and fishery management. *J. Cons. CIEM*, **33** (3): 471–7.

Gulland, J. A. (1974) *Guidelines for Fisheries Management.* FAO, Rome. IOFC/DEV/74/36: 84 pp.

Paulik, G. J. (1971) Review of J. A. Gulland, *Manual of Methods of Fish Stock Assessment. J. Cons. CIEM*, **34** (1): 137–139.

Paulik, G. J. (1972) Digital simulation modeling in resource management and the training of applied ecologists. In *Systems Analysis and Simulation in Ecology*, G. Patten (ed.). New York, Academic Press.

Regier, H. A. (1976) Environmental biology of fishes: emerging science. *Environ. Biol. Fish.*, **1** (1): 5–11.

Roedel, P. (ed.) (1975) Optimum sustainable yield as a concept in fishery management. *Spec. Publ. Am. Fish. Soc.*, (9): 89 pp.

Rothschild, B. J. (ed.) (1972) *World Fisheries Policy: Multidisciplinary Views*. Seattle, University of Washington Press, 272 pp.
Zalinge, N. van and Nurzali Naamin (1975) The Cilacap based trawl fishery for shrimp along the south coast of Java. *Laporan Penelitian Perikanan Laut* (*Mar. Fish. Res. Rep. Jakarta*), **1975** (2): 1–44.

ADDITIONAL READING (outside fishery literature)

Gold, H. J. (1977) *Mathematical Modelling of Biological Systems—An Introductory Guidebook*. New York, Wiley–Interscience.
Krebs, C. J. (1978) *Ecology: The Experimental Analysis of Distribution and Abundance*, 2nd edn. New York, Harper and Row.
Patten, B. C. (ed.) (1971) *Systems Analysis and Simulation in Ecology*. New York, Academic Press, 4 vols.
Snedecor, G. W. and W. G. Cochran (1967) *Statistical Methods applied to Experiments in Agriculture and Biology*, 6th edn. Ames, Iowa State College Press.

EXERCISES

Exercise 1.1

The catch (in tons) and effort (number of boat-months) in the shrimp fishery at Cilacap, southern Java, are given in Table 1.3

Table 1.3

	1971	1972	1973	1974	1975
Catch (tons)	170	3800	4500	3000	3050
Effort (boat-months)	48	350	1200	1400	1450

(Data adapted from van Zalinge and Nurzali (1975), to correspond to average conditions of rainfall and other environmental factors.)

Determine graphically the relation between catch and number of boat-months (assuming for the purpose of this exercise that each year's data corresponds to an equilibrium position). Is it likely that there was an increase in efficiency (due to learning) between 1971 and 1972? What is the MSY, and how many trawlers are needed to take it?

If 1 kg of shrimp is worth 500 rupiah, and it costs 900 000 rupiah a month to operate a trawler, at what size of fleet will costs equal value of catch? (Convert catch and effort to rupiahs and replot the data, or draw lines of equal costs and values on graph of tons caught against number of trawlers.) How does this compare with the recent situation of the fishery? What is the MEY and the net economic return at the MSY? (For MEY, find the point on the yield curve where the tangent is parallel to the line of equal costs and values.)

If each trawler has a crew of six men and the current situation of unemployment is such that the government is prepared to sacrifice a net income of 1 million rupiah to ensure employment of one man which, on this basis, is the more desirable state for the fishery (a) at MEY, (b) at MSY, (c) with a fleet of 100 trawlers working 12 months a year? (Calculate employment and net economic yield for each, and the differences between them.)

CHAPTER 2

Basic concepts and data sources

In the preceding introductory chapter an outline has been given of what stock assessment is about, and the purposes it tries to serve. The following chapters describe the techniques used in achieving these purposes. These techniques are concerned with the manipulations of certain types of basic information (the total weight caught, numbers of fish of a given age in a certain sample, etc.). Before describing these manipulations it is desirable to describe at least briefly what information is needed and how it is obtained, but even before that it is necessary to discuss the framework and limits within which the manipulations are made.

2.1 THE UNIT STOCK

The models discussed in the following sections treat the fish populations as though they were uniform. The analytic models do distinguish individuals according to their sex, age, or size, but otherwise assume that each individual of a given size or age behaves in the same way. This is a necessary simplification in order to make the models workable. In fact fish differ; feeding may be better in one part of the range, resulting in a better growth; fishing is not distributed uniformly, so that some fish are exposed to a greater intensity of fishing than others. Within a small area there may be sufficient and rapid mixing so that over a period it is reasonable to suppose that individuals are exposed to something approaching average conditions of feeding, or fishing intensity, and differences within the area can be ignored. For larger areas this may not be so reasonable; differences between different parts can be marked and consistent, amounting sometimes to clear genetic differences within a single species.

It can be a matter of some importance to choose an appropriate group of fish, the unit stock, that can be treated as a homogeneous and independent unit. Too large, and important differences within the unit stock may be neglected; too small, and interactions with other groups of fish may be important and the need to consider these interactions may add considerably to the complexities of analysis.

For many purposes of stock assessment analysis, and the use of that analysis in providing advice, the choice and definition of a unit stock can be considered as essentially an operational matter, being tied to the models being used, the questions being asked, and the volume and detail of information available. Briefly, a group of fish can be treated as a unit stock if possible differences within the group and interchanges with other groups can be ignored without making the conclusions reached depart from reality to an unacceptable extent.

The ideal stock is that described by Cushing (1968) as one that 'has a single spawning ground to which the adults return year after year. It is contained within one or more current systems used by the stock to maintain it in the same geographical area'. Some unit stocks, e.g. the Arcto-Norwegian cod, fit this well and sometimes (again the cod fits) the scientist is fortunate that a fishery (or a group of fisheries) is directed almost wholly on to a single distinct unit stock. This is not always the case. The question of defining a unit stock has received most attention in respect of the Pacific salmons (e.g. Ricker, 1972) in which the ability of the adult fish to return to a distinct spawning stream has probably facilitated the genetic separation of stocks and races, and has certainly made the existence of clearly separated stocks, each requiring its own harvesting strategy for optimum results, of great practical importance. Salmon are probably in this (and other matters) exceptional.

A unit stock may correspond to a distinct taxonomic unit—a species or subspecies. For example, the southern bluefin tuna is distributed over the southern subtropical and temperate oceans from New Zealand to southern Africa. Tagging has shown movements of fish over the entire range and these fish probably form a fairly homogeneous group, spawning in the subtropical eastern Indian Ocean, spending their first few years of life off the Australian coast, and the adult fish feeding in the cooler waters of the southern Indian Ocean. This species can be treated as a unit stock.

Clearly, if a taxonomic group is treated as a unit stock, there are no problems of mixing with fish outside the group, and the group is not, on these grounds, too small but it may be too large. Within a taxonomic group there may be subgroups that differ quite substantially in their population characteristics (e.g. growth), or in the impact of fishery on them. A simple situation occurs when there are discontinuities in the distribution. For example, though the distribution of the North Atlantic cod includes all shelf areas in an arc from New England through southern Greenland to the English Channel, there are troughs of deep water which cod do not frequent and across which there is little movement. The deep-water zones enable certain stocks, for example at Iceland and around the Faroe Islands, to be fairly clearly identified as unit stocks.

There need not be a break in the distribution of a species for differences to be significant, and for different groups to be treated as separate unit stocks. For example, the yellowfin tuna occurs right across the tropical Pacific, but individuals do not seem to move over such long distances as other tuna species. Thus the fish in the east, off the American coast, which are heavily exploited by the surface fishery (now mainly carried out by purse-seiners), are treated as a unit stock, independent of the fish in the central and western Pacific, exploited mainly by long-liners. Further separation within the eastern Pacific fishery may also exist.

Since the choice of a unit stock is largely subjective and determined by operational considerations—it *is* a unit stock if it can be treated as a unit stock—the size of the group treated as a unit stock may well change if conditions change. For example, unless mixing is rapid throughout the group of fish considered, it is desirable, if it is to be considered as a unit stock, that fishing is fairly uniform throughout the distribution of the group considered. Therefore the distribution of fishing is important in determining what can be considered as a unit stock. This can easily change. For example, the distribution of the surface fishery for yellowfin tuna in the eastern Pacific has extended westward following the change from live-bait fishing to purse-seining around 1960, and this extension has received a further impetus from the regulations of I-ATTC. As a result the group of fish that need to be considered as the 'unit stock' when analysing the surface fishery data has grown. This explains to a large extent why the estimates of the potential yield from the current fishery (see reports of Inter. American Tropical Tuna Commission, I-ATTC) are well above the early estimates of Schaéfer (1957).

The group of fish treated as a unit for assessment purposes need not necessarily consist only of fish of the same species. Fish of different species clearly will differ to some extent in their characteristics (growth, natural and fishing mortality, etc.) and an analysis that ignores these differences cannot be entirely correct. On the other hand the differences may not be great, and the alternative of treating each species separately may be impracticable. In many tropical trawl fisheries several dozen species may be taken in a single haul, and 100 or more may make significant contributions to the fishery. The fishermen cannot concentrate their attention on individual species, but must give their attention to the complex of bottom-living fish considered as a whole. Treating the population of bottom fish as a unit resource therefore give quite reliable results, as well as being the only practicable approach.

2.1.1 Determination of a unit stock

Little more than passing attention is normally given in the early stages of an investigation to the question of what constitutes a unit stock. A common course is to take the data referring to the fishery of direct concern—the English fishery for plaice in the North Sea, the anchovy fishery of Peru—and proceed directly to the analyses of major interest (estimating growth and mortality, plotting catch per unit effort against average effort, etc.). The implicit assumption is that the fish do effectively constitute a unit stock. If possible this assumption should be checked.

The two ways in which the assumption can break down—because the fishery is based on several independent stocks or because it exploits only part of a large stock—have different consequences, and need checking in different ways. If there are several different stocks, the danger is that each stock will be behaving in a different way. The fishery will produce a picture that is in some way a weighted average of the events in each stock. If this averaging is consistent from year to year, then it may be that different behaviour of unit stocks within the fishery will not seriously invalidate any analysis of how the stocks as a whole, averaged together in the way that is done by the fishery, will behave. That is, so far as most of the immediate applications of stock assessment studies—advising on the change in total catch to be expected from an increase in the amount of fishing, etc.—are concerned, the possible existence of separate stocks is not serious so long as the distribution of fishing between the stocks is unchanged.

The most serious change is an expansion of the range of the fishery such that new stocks become exploited. This is common in coastal fisheries based on a series of local independent stocks along the coast (e.g. shrimps). These may start on a single stock near the main harbour; as this becomes heavily exploited the fishery spreads to other stocks further along the coast. In this way the catch per unit effort may be maintained, or at least decrease only slightly, and therefore a superficial analysis may suggest that fishing is having little effect on the stocks.

The first step should, therefore, be to examine the distribution of fishing (in both space and time, since seasonal migrations can result in different unit stocks inhabiting a given area at different times of the year). Particular attention should be given to any expansions in the area fished, and to the possibility of expansion in the future.

If the pattern of fishing has changed, or may change in the future, a number of aspects can be examined to provide information on possible stock separation. These include:

(a) Distribution of fishing. A gap in fishing suggests a gap in the distribution of fish, which may correspond to a separation of stocks.
(b) Spawning areas. A genetic separation of stocks more or less requires a clear separation of spawning groups, even if the fish mix at other times of life. Spawning grounds may be delimited by detailed surveys for spawning fish, or for eggs or the youngest larval stages, but may also be determined quite well from information obtained from commercial fishermen.
(c) Values of population parameters. If there are stock differences, and if these differences are important, then there should be differences in some at least of the population parameters (growth, mortality, etc.). However, within a unit stock there may be large and systematic differences in age or size composition between various grounds. Average age (or size) is therefore a poor parameter to use. Better quantities to look at are estimates of total mortality (and especially changes in this), growth or the occurrence of good or bad year-classes. Better are changes in these. An increase in mortality in one area, combined with a lack of change in another is good evidence of stock separation, particularly if the change and lack of change in mortality are consistent with changes in the amount of fishing in the two areas.
(d) Morphological or physiological characteristics. Characteristics that are genetically determined (e.g. Cushing, 1964) can provide clear evidence that two groups are distinct, but genetic separation can in principle exist without this being evident in the characteristics examined. Further, mixing can proceed fast enough to achieve genetic uniformity while still being slow enough for different groups to be separate stocks for practical assessment

purposes. Other characteristics that have been extensively used to determine stock structure, for example vertebral counts, are not determined wholly by genetic factors; for example, temperature at hatching may help determine the number of vertebrae. A difference is therefore not clear evidence of genetic separations, and may also exist when mixing is sufficiently rapid to make the fish effectively a single stock.

(e) Tagging. In principle this can give the clearest evidence of stock separation or otherwise. If fish tagged on one fishing ground are later caught on another, then the same stock occurs on both grounds (though other stocks might also occur); if not, then the stocks are separate. While the existence of mixing is clearly shown by interchange of tagged fish, the failure of tagged fish to turn up on the second ground is not conclusive evidence of separation. The exploitation rate, or the efficiency with which the fishermen detect and return fish, can vary between areas, and the absence of returns may mean the fish are there, but are not caught or, if caught, the tags are not returned. Also, returns of tagged fish are often concentrated soon after the time of release, when they have not had time to move far. A better impression of the degree of mixing, and the separation or otherwise is obtained by discarding the early returns from most considerations. By far the best way of determining from tagging the degree of interchange or separation between two fishing grounds is to tag on both grounds, if possible simultaneously.

When the bounds of the unit stock extend beyond the limits of the fishery being analysed, the effects again depend to a large extent on the amount of change in the pattern of fishing. If there are other fisheries exploiting the same stock and they develop with the same pattern and timing as the fishery being analysed, then the results of the analysis may not be misleading. The scientists may for example be analysing half the total amount of fishing on half the total stock, and thus obtain the correct answer. Independent fisheries in the same region may well develop in a similar pattern in response to similar economic, technological, or social factors, but this need not necessarily be the case. The possibility of the stock being affected by events outside the fishery of direct interest must always be considered.

The techniques—morphometrics, tagging, blood-typing, etc.—which are useful for determining whether groups of fish within the same fishery form a single stock or a number of distinct stocks are also useful for determining what other fisheries, if any, in addition to the one of prime interest act on the same unit stock. In addition consideration of the migration pattern can help. The problems of determining what other fisheries are exploiting the same stock will be greatest when the species concerned makes long migrations. The study of fish migrations is beyond the scope of this manual (but see for instance Harden-Jones, 1968), but is likely to be crucial in understanding the stock structure of widely ranging fish. An examination of when and at what ages (or sizes) fish occur in different fisheries can suggest which fisheries could be exploiting the same unit stock at different stages in its migratory cycle. The existence of a gap in the seasonal fisheries, or in the sizes caught, can also be of direct practical importance in suggesting possible extensions of the current fisheries to new areas.

The extension of national jurisdiction over fisheries—usually to 200 n.m.—as a result of changes in the accepted law of the sea has brought the question of the identity of unit stocks into sharp focus. If one country is interested in developing or managing the fish in its area of jurisdiction, it needs to know to what extent its actions will affect, or be affected by, actions taken by adjacent countries. Two classes of stock can then be distinguished—those that occur wholly within the area of jurisdiction of a single country, and those that are shared by two or more countries. A stock in this sense involves slightly different concepts than those of a unit stock in the sense used in the preceding paragraphs. A 'shared stock' may be, but need not necessarily be, a 'unit stock'. For example, the Arcto-Norwegian stock of cod has a single well-defined spawning ground of the Lofoten Islands, and a pattern of migration that brings many fish into the waters off northern Russia. These cod form a unit stock that is shared by Norway and the Soviet Union. Its assessment involves the combination of data from both countries (and from any other country fishing it), but the questions of national jurisdiction raise no special problems. It makes little difference to the assessments where a particular fish is caught—the important difference in the sizes of fish caught in different zones should be taken into account in the normal assessment procedures.

In contrast there are other species, for example many demersal species in the Gulf of Guinea, that do not have well-defined spawning grounds, nor clear migration patterns. They do move however, and do not respect man-made boundaries. The fish off, say Ghana, are therefore not independent of those off the Ivory Coast, and thus in one sense form a shared stock, and it is not possible to treat the fish off Ghana and the fish off the Ivory Coast as two separate unit stocks. At the same time, depending on the species, the degree of movement may be insufficient for the interaction to be very important except in the border areas of the two countries, and it might be equally misleading to treat the fish off both countries as a single unit stock. The full assessment of fisheries of this type will involve a study of the migration and dispersion patterns of the fish, and the incorporation of suitable expressions for the interchange between areas into the population models. This is outside the scope of the present manual. In the absence of suitable models it will often be useful to carry out two different sets of assessments, based on the hypothesis of a single unit stock, and of two independent stocks. The management of shared stocks is discussed further by Gulland (1980).

2.2 DATA FROM THE COMMERCIAL FISHERY

The commercial fisheries—which in this sense include also small-scale and artisanal fisheries—are both the final users of stock assessment studies and the source of some of the most important data used in these studies. Close links with the fisheries are therefore most important for effective stock assessment work. The routine supply of data from these fisheries has been described in detail elsewhere, including other FAO manuals (see especially Brander, 1975, for a general outline, Holden and Raitt, 1974, Anon, 1981, for the collection of biological data, Bazigos, 1974, for statistical surveys and several fishery circulars issued by the FAO Fishery Statistics Unit is respect of the standards and classifications used in regional and international statistics), but in view of the importance and relevance to this manual, the main points are worth summarizing here.

2.2.1 Types of data

The main types of data coming from the commercial fisheries concern:

(i) total catch;
(ii) amount of fishing;
(iii) catch per unit effort;
(iv) biological characteristics of the catch (size, age, etc.) each of which present particular problems.

The total catch is the most vital information, without which any assessment studies are meaningless. It may be possible, for example, to make reasonable first approximations to the potential yield that can be taken from a body of water, using data on its physical and biological characteristics relative to those of other better-known waters (e.g. Ryder, 1965, Henderson *et al.*, 1973) but this does not tell much about the state of exploitation unless the present yield is known and can be compared with the estimated potential.

Catch data should include all removals from the stock by all the fisheries on it. This means checking on the extent of the unit stock, as discussed in section 2.1, and ensuring that none of the minor fisheries (including subsistence or sports fisheries) are omitted. Some care is needed in interpreting data from all but the simplest fisheries. After capture the fish caught are usually sorted; the least valuable (small, or of undesirable species) may be discarded; the rest of the catch may be gutted and cleaned, the heads may be removed; in quite primitive tropical fisheries the catch may be dried; and in the most modern fisheries the fish may be filleted and frozen. The weight of the fish, as they are put ashore from the fishing vessel (or from the mother ship or carrier vessel) may be less, and possibly much less than the weight as they leave the water. The former—the landed weight—is the quantity usually recorded in original records. If the analysis of what is happening to the stocks is not to be confused by differences in the ways in which fishermen sort and prepare the fish at sea, it is important to express the landed weight in standard terms. This is normally done by expressing all catch and landing statistics in terms of the weight of the fish as they were removed from the water—live weight or round fresh weight. Usually discards are omitted from consideration because information on them is very scarce; the statistics are in that case strictly of 'nominal catch'. However, if the full impact of fishing on the stocks is to be accurately assesed, some estimate of the discards should be obtained.

Data on the amount of fishing are used for two distinct purposes: in stock assessment (in a fairly narrow sense), as a measure of the fishing mortality coefficient (see Chapter 4), and in general economic studies, as a measure of the costs of fishing. The same series of basic data, e.g. the number of days spent at sea by vessels of a fleet each year, may be used for both purposes. In the short run, and to a first approximation, both may be reasonably well served, but in the long run, to take account of technological innovations or to adjust the basic statistics to reflect more accurately the quantity of real interest, each purpose will require different treatment. For instance, suppose the introduction of sonar doubles the fishing power of a purse-seiner, i.e. doubles the catch it will take per day fishing on a given density of fish. From the point of view of a biologist the fishing mortality caused by day fishing has doubled, and the number of days fishing with sonar should be multiplied by two to give the fishing effort in nominal numbers of pre-sonar days. To the economist the input has increased only slightly (by the cost of hiring or purchasing the equipment) and little adjustment is necessary. The present manual is mainly concerned with the biological viewpoint of fishing effort, but the types of information collected from the fishery—the number of vessels, their main characteristics (tonnage, horsepower), the type of fishing gear and other equipment, and the time spent fishing and number of operations (trawl hauls, purse-seine sets, etc.) are in their basic form much the same.

In principle, just as the scientist must take account of all the catches in the stock, he must also take account of all fishing on the same stock in calculating the amount of fishing, and data on fishing effort from all components of a fishery are desirable. This is often not easy, and in practice total effort is usually estimated from data on a part only of the fishery.

Catch per unit effort data are used to estimate changes in stock abundance. Since it is often difficult or costly to estimate the abundance in other ways (e.g. by surveys with research vessels), and some measure of abundance or at the least, of changes in abundance, is vital in any stock assessment study, obtaining reliable catch pre unit effort data is (except when survey or similar data are available) one of the most important basic steps in such study.

There will be occasions when it may be suspected that the available statistics of catch per unit effort (c.p.u.e.) data do not give a reliable index of abundance (for example, in a purse-seine fishery where the success or otherwise of searching for schools determines the size of the catch). However, until some other and better measure of abundance is available, the only alternative to doing nothing is to use the c.p.u.e. data, though with suitable precautions about the sort of errors in the final conclusions that might be introduced by the ways in which c.p.u.e. figures misrepresent the abundance, and with maximum attention being given to ways of obtaining better units of effort and c.p.u.e. Catch per unit effort data is only necessary from whatever part of the whole fishery is most likely to provide a good index of abundance, though the larger the part the smaller is likely to be the random variation. The basic information collected from the fishery will naturally be the same as that in recording the separate statistics for estimating total catch and total fishing effort. The problems of obtaining good estimates of total effort and of c.p.u.e. are in fact completely intermixed and are discussed together in more detail in section 2.3.

The commerical fisheries are also invaluable sources of biological data on the fish stocks, notably on the sizes and ages of the fish. Information on the fish in the catch are of interest in themselves (e.g. cohort analysis, see section 4.3.5, depends on having information on the total numbers of each age caught), but also in providing information on the age of size composition of the stock. All fishing gear is selective, and commercial catches will be further selected by the fisherman choosing to fish at times and places where the most abundant sizes (or the sizes with the most favourable combination of abundance and market price) are present. Nevertheless, the commercial catches usually do provide a useful estimate of the composition of the stock, at least over a reasonably wide range of sizes or ages. Since it is usually necessary to combine information from all fisheries on the same stock—often from different countries—some standardization is desirable. For example, it is useful for all scientists studying the same stock to measure the same characteristic (e.g. total length or fork length) and group measurements in the same way (e.g. by 1 or 5 cm groups).

2.2.2 Collection of the data

Where data of general interest (e.g. total catch, number and size of trawlers) are being collected from large-scale industrial fisheries, complete records may be easily acquired. For example, data on the number of larger vessels and also on their size and other characteristics will probably be available from some form of vessel registration; at the bigger ports the port authorities may have records of arrivals and departures from which the number of trips and the length of each can be derived; commercial fishing companies will have good records of the total weight and of the main varieties of fish landed—though because of taxes accurate records may not always be available to government authorities. Where the number of landing places are few, a small number of collectors of fishery statistics can provide complete records of several types of data—number of landings and catches.

Complete records cannot be obtained on other situations. It is only possible to measure a few of the fish landed by a proportion of the total vessel landings or to make direct records of the catches of a few of the thousands of scattered subsistence fishermen in Asia or of sports fishermen in more developed countries. For these data some form of sampling system is necessary. These are well developed in agricultural statistics, and are becoming more generally used in fisheries (Bazigos, 1974). The use of sampling methods to obtain catch statistics has sometimes been held up by a feeling of biologists or administrators that the data obtained are necessarily inferior to data from complete records. This is not necessarily so, and sometimes because the original observations are made by special records, the sampling data may be more reliable. They will be subject to sampling variation, but this can be determined, whereas the errors in other types of data are normally unknown, even when they may be considerable. It is important, when setting up a sampling system, to reduce sampling variation by suitable design of the system, and also to determine what precision is needed. The costs of a sampling survey increase rapidly as greater precision is needed. Conversely, if only moderate precision—say $\pm$ 10 per cent in the estimate of the total weight landed is needed—this can often be achieved very cheaply.

2.3 EFFORT AND CATCH PER UNIT EFFORT

2.3.1 General considerations

In discussing catch, effort, and c.p.u.e., the scientist is concerned with three closely related characteristics of the fish stock: the quantity removed from it by fishing C; the rate at which fish are removed, or fishing mortality F; and the abundance N. (The symbols for catch and stock abundance used here refer to numbers, but the same relations are true for weight and biomass.)

These are connected as instantaneous rates by the relation

$$\frac{dC}{dt} = FN$$

or, over a unit time period, say a year, by

$$C = F\bar{N} \tag{2.1}$$

where $\bar{N}$ is the average abundance during the year. Clearly, if any two of the quantities in equation (2.1) are known, the other can be immediately calculated.

Normally the absolute values of neither the abundance nor the rate of fishing (the fishing mortality) will be known, but estimates will be available of quantities that can be used as indices of these population characteristics. These are the effort f and the c.p.u.e. $\bar{U}$. These will be related to the mortality and abundance by the equations

$$F = qf \tag{2.2}$$

and

$$\bar{N} = \left(\frac{1}{q}\right)\bar{U} \tag{2.3}$$

where q is the catchability coefficient.

The catchability coefficient is presumed to be constant. The value and reliability of a given set of statistics of effort and catch per unit depend on how closely this presumption is satisfied.

Sometimes the scientists investigating a given stock have available, or can arrange to have made available, independent indices of abundance of a stock (e.g. from acoustic surveys) (see section 2.3.6). In that case assessment of the stock can proceed (assuming catch data are available) in terms of abundance (in relative or absolute terms), and fishing mortality.

More often the only measures of fishing mortality or abundance that are available are those derived from information on effort or catch per unit effort. If any but the simplest assessment of the stock is to be made, some effort data will have to be used. The question is then which effort (or c.p.u.e.) units are most satisfactory, how could they be adjusted to make them more satisfactory, and if no completely satisfactory adjustment can be made, what errors might be introduced into the assessments. This is equivalent to examining the possible extent of variation in the catchability coefficient q, and finding the unit of effort (or c.p.u.e.) for which the value of q is most nearly constant, taking account of any adjustment that can be made to the effort units.

The nature of the changes, if any, in the value of q can be studied from a number of aspects—in terms of the type of change (cyclical, trends, correlated or not with abundance or amount of fishing), in terms of the cause of the change (for example increased power of fishing vessels), or in relation to the type of fishing gear used.

The test of whether the units used are good, i.e., q is constant, can be made by looking at effort and mortality, or c.p.u.e. and abundance. That is, either one can consider whether a given amount of effort will always remove the same proportion of the stock, and changing the effort by a given percentage will change the rate of removals by the same percentage, or one can consider whether a given c.p.u.e. will always correspond to the same abundance, and a change in abundance of a given percentage will be reflected by a change in c.p.u.e. of the same percentage. The two approaches are equivalent, but the abundance of a fish stock is a concrete concept, and more easily considered than the abstract concept of fishing mortality; the second approach (considering whether or not c.p.u.e. is a good measure of abundance) may be found easier.

The examination of the interrelation between effort, mortality, c.p.u.e., and abundance is concerned almost entirely with *changes* in the catchability coefficient q. No statistics of effort provide directly and immediately absolute values of mortality (e.g. as the percentage of the stock caught per day or per year), nor do any c.p.u.e. data provide absolute values of abundance, in terms of numbers or weight of fish. Therefore, a given unit of effort or c.p.u.e. cannot properly be said to be biased, in the sense that they give estimates (of mortality or abundance) that are consistently too high or too low. Effort units give biased estimates of the *change* in effort, or c.p.u.e. when there is a trend in catchability in time, e.g. increasing, such that in later years the effort statistics underestimate the mortality (and c.p.u.e. overestimate abundance) relative to the indices of mortality (or abundance) obtained from the effort or c.p.u.e. data in earlier years.

It is also true, because effort and c.p.u.e. provide only indices of the quantities of real interest, that there is no such thing as the single correct unit of effort or c.p.u.e. for a given type of fishing gear. Any unit will be valid in respect to a particular analysis if, over the space and time covered by that analysis, the catchability coefficient q corresponding to that effort unit remains constant. A number of different units may be valid for the same period, and generally it is most convenient to use the simplest one. For example, the fishing power of trawlers increase roughly in proportion to their tonnage, so that expressing the fishing effort as the sum over the whole fleet of the product of the gross tonnage of each vessel times the number of days spent fishing is likely to provide a better measure of effort (i.e. one for which the catchability changes less) than the simple number of days fishing. However, for a homogeneous fleet of 100 ton vessels the two sets of effort data (days and ton-days) will be identical except for two zeros, and there is no advantage of one unit over another, at least until such time as some smaller or larger vessels join the fleet.

This example shows that the choice between effort (or c.p.u.e.) units is not absolute, and the preferred unit, and the adjustments that are made to it, may change as the fishery changes. In some cases the choice between two alternative units may be reverse. For example in the English distant-water trawl fleet there are two measures of fishing time that have been used—the number of days at sea, and the number of hours spent fishing, i.e. with the trawl actually on the bottom. At one time, when fish were abundant, fishing was interrupted to allow the crew time to gut, clean, and stow the fish. A day at sea then represented a variable fishing mortality, depending on the number of interruptions, which in turn depended on the fish abundance. The catch per day was determined more by the speed of work of the crew than on the abundance of the stock. The number of hours fishing was at that time a better unit of fishing effort. More recently, as trawlers have been able to search actively with advanced acoustic gear for worthwhile contributions of fish, the number of hours fishing per day has varied because trawlers tend to shoot their net only when they have found worthwhile concentrations. The catch per hour, therefore, now mainly reflects what is considered a worthwhile concentration. The stock abundance is better measured by the time taken to find good concentrations, i.e. by the catch per day. The number of days at sea may now be a better unit of fishing effort than hours of fishing.

The choice of fishing effort unit must therefore be based on a close understanding of how the fishery operates, and what changes in the fishery—the vessels, the gear they use, the strategy of the fishermen, or in the stock itself—might be occurring which would change the catchability coefficient applicable to the effort unit being considered. There cannot be an exact formula for choosing the best unit for a particular fishery, but general guidance can be found in the following section.

2.3.2 Patterns of changes in catchability

Catchability is in practice far from constant if the fishery is looked at in sufficient detail. Catches from a single haul, even by the same vessel at the same place and at nearly the same time, can vary greatly. The variation naturally increases when considering different vessels, different parts of the area inhabited by the stock, and different times and seasons. Fortunately, much of this variation affects the assessment of the stocks to only a minor extent. The degree of impact on assessment depends on the patterns of change in the catchability. These can be classified as follows:

Cyclical changes in time (according to day, season, etc.).
Trends in time.
Changes related to stock abundance.
Changes related to the total amount of fishing.
Random variations.

Cyclical and systematic changes in catchability, for example, catches of some species being higher at night than during the day, or the large catches often taken on pre-spawning concentrations are of great interest to the fishermen, but need not be of much concern in stock assessment. For assessment, the greatest interest is in year-to-year changes. The catchability coefficient applicable to the annual values of effort or c.p.u.e. will be the average, taken over one or more complete seasonal or other cycles. These will, so far as the cyclical effects are concerned, be the same from year to year, *provided* the averages are taken each year in the same way. In effect the catchability coefficient $\bar{q}$ corresponding to the annual figures of catch and effort is the weighted average of the seasonal catchability coefficient, with the weighting factors being equal to the fishing effort in each season. That is, putting $\bar{N}$ = average abundance during the year, and ignoring seasonal changes in abundance (these can be taken account of without changing the conclusions), for the whole year, $C = \bar{q}f\bar{N}$ and for the ith season (or other time period)

$$C_i = q_i f_i \bar{N}$$
$$C = \sum C_i$$

and

$$f = \sum f_i$$

hence

$$\bar{q} = C/f\bar{N} = \frac{\sum (q_i f_i \bar{N})}{\sum f_i \bar{N}} = \frac{\sum f_i q_i}{\sum f_i} \tag{2.4}$$

Changes in the distribution of fishing effort between seasons or other periods can therefore change the value of the effective catchability corresponding to the use of total annual effort or c.p.u.e. for the whole year. For example, proportionally more fishing at times when fishing is good will increase the catch and c.p.u.e. from a given stock abundance and the same total effort. The fishery must therefore be carefully monitored to detect if changes in fishing practice are occurring.

If data are available for each period (season, month, etc.) separately, then adjustment for changes in the way in which the fishery distributes its efforts between different periods is possible, and should generally be made. The c.p.u.e., U_i for each period can be calculated separately, and the average of these $\bar{U}$ will give a c.p.u.e. for the year, for which the catchability coefficient may be noted by q'. Then, if n = number of periods,

$$U_i = C_i/f_i = q_i\bar{N}$$

$$\bar{U} = \frac{1}{n}\sum U_i = q'\bar{N} \tag{2.5}$$

and hence

$$q' = \frac{1}{n}\sum q_i$$

This shows that the value of q' is not affected by the pattern of fishing, i.e. the way in which the total annual fishing effort is distributed between seasons. The corresponding measure of the total effort for the year, f' is given by

$$f' = C/\bar{U} \tag{2.6}$$

By expanding equation (2.6) in the form

$$f' = \frac{1}{\bar{U}}\sum C_i = \frac{1}{q'\bar{N}} \times \sum (q_i f_i \bar{N})$$

$$= \frac{n\sum q_i f_i}{\sum q_i}$$

it can be seen that the effective amount of fishing, i.e. the value of f', depends on the seasonal distribution of the effort, as well as the total nominal effort Σf. This is obviously reasonable.

When information from a number of years is available, the same data can be presented in a slightly different form. The c.p.u.e. for a given period, for example July 1975, can be expressed as a percentage of the average, over some standard set of years, of the c.p.u.e. in July. The resulting figures can be treated as standardized measures of c.p.u.e., and the monthly values averaged to give an annual c.p.u.e. The year-to-year changes in this c.p.u.e. will be the same as in the direct average of the simple monthly c.p.u.e., but the array (by months and years) of the standardized values make it easier to see how consistent (or inconsistent) are the seasonal patterns.

Different adjustments are necessary for other changes in the temporal pattern of fishing effort for which it is not possible to collect data for separate periods. For example, when the catches at night are much lower than during the day, fishermen often only fish during the day, but may change these habits and also fish at night if demand for fish is very high. For a given abundance of fish such a change will increase the catch per day (since more fishing will be done), but reduce the catch per hour of fishing (since more fishing will be done when catches are relatively poor). Suitable adjustments to the statistics of fishing effort for the period after the change would therefore be to increase the nominal number of days by a suitable factor, or to reduce the number of hours by some other factor. These factors would have to be estimated from special studies on the relative catch per day, or catch per hour, obtained by vessels using the two fishing practices.

Trends in catchability over periods of years are the commonest reasons for having to adjust the effort units being used. They are generally caused by increases in the efficiency of the fishery, so that a unit effort removes a greater proportion of the stock, and the c.p.u.e. corresponding to a given stock abundance increases. The possible changes and methods for dealing with them are discussed in more detail in section 2.3.3, in relation to the causes (larger or more powerful vessels, improved gear, better searching, etc.) Here, it should only be emphasized again that the main concern is with *changes* in the fishery (whether in the size or power of the ships, in the actual fishing gear, or in the fishing tactics, and the auxiliary equipment used). If, in the extreme example, no changes take place in vessels or gear, or in the way they are operated, then no adjustments are needed, and any available unit of effort (number of boats, number of fishermen, etc.) can be used. Usually, any successful group of fishermen will be continually changing their practice and improving their performance. The effort units used must be kept under continuous review to ensure that the corresponding catchability is not being increased by those improvements.

Changes related to *stock abundance* can be the most serious types of change in catchability, since they can partly or wholly obscure what is happening to the abundance of the stock. In the extreme case, if catchability is inversely proportional to abundance, then

$$U = q\bar{N}$$

and

$$q = a/\bar{N}$$

therefore

$$U = a$$

i.e. the c.p.u.e. is constant.

This is not a wholly fanciful supposition. In a purse-seine fishery, the catch per shot of the net might appear to be a suitable measure of c.p.u.e. If, however, as the stock decreases the number of schools decrease, but the size of the school remains constant, and each shot takes one complete school, then clearly the c.p.u.e. is equal to the school size, and constant. The catchability, i.e. the proportion of the total stock taken by a unit of effort (one shot), is inversely proportional to the number of schools.

Changes in stock abundance can affect catchability in several ways. The simplest is when fishing is a continuous process, and the size of the catch taken early in the operation can affect the later efficiency of the gear—often referred to as gear saturation. An obvious example is a long-line where the fishing power is presumably proportional to the number of unoccupied hooks. For a long-line, with n hooks, it can be shown (e.g. Gulland, 1955, equation 2.10) that the number of fish caught, C, after the gear has been fishing for time t, is given by

$$C = n(1 - e^{-aNt}) \tag{2.7}$$

where N = number of fish in the population, and a = probability that a particular fish will be caught by a particular hook in unit time.

If aNt, and hence the proportion p of hooks that are occupied by time t is small, equation (2.7) reduces to $C = naNt$, i.e. the catch is proportional to the stock abundance. As N and hence p increases, then C increases less than proportionally to the abundance, i.e. the c.p.u.e., if expressed in such simple terms as catch per 100 hooks, will not be a satisfactory measure of abundance. Under certain assumptions (e.g. that fish encounter the gear independently), a satisfactory measure of effort will be

$$f' = -\frac{np}{\log(1-p)} \tag{2.8}$$

where n = number of hooks, with the corresponding measure of c.p.u.e.

$$U' = -\frac{C\log(1-p)}{np} \tag{2.9}$$

(the minus sign occurs because $\log(1-p)$ is negative).

Quantitative adjustments, using equations (2.8) and (2.9) may only be possible in exceptional circumstances, for example a long-line fishery, in which the quantitative reduction in the effectiveness of the gear can be estimated. For such fisheries it must be noted that the reduction in effectiveness is caused by any fish occupying the hook, and not merely fish of the stock of immediate interest. In other fisheries, such as gill-nets and lobster pots, it may be clear that early catches will reduce the effectiveness of the gear, but not by how much. Quantitative correction may be difficult, though the direction of the errors introduced, i.e. that the c.p.u.e will understate changes in abundance, will be known.

Changes in abundance can also affect the catchability corresponding to a given unit of effort by changing the proportion of the total recorded time that is actually significant in catching fish. An example of this concerning the English distant-water trawl fleet has been given earlier, where the amount of time that the net was actually on the bottom fishing fell off as abundance increased. This is a form of 'saturation' in which it is not the gear itself that becomes saturated but the operational performance of the vessel. Provided the appropriate unit of fishing time can be defined and recorded, there is no problem, once the possible difficulty has been recognized.

A particular subtle form of this type of saturation concerns gear (e.g. purse-seine) in which the critical part of the fishing operation, which determines what proportion of the stock is taken, is not the actual process of catching (e.g. setting and hauling the net) but the preceding searching time, in which the school on which the net is set is located. For schooling fish, which is the object of these fisheries, the average catch per successful shot will, if the entire school is taken, be equal to the average school size, and the total number of schools will—provided the area in which schools may occur does not change with changes in abundance—be proportional to the number encountered per unit searching time. For these fisheries good effort data depends on breaking down the total time available from simple records (e.g. number of trips or days at sea) into searching time and other time (see section 2.3.3). This may be difficult both practically (because it may need individual fishermen to keep detailed records), and theoretically (which searching time, if any, is spent by a vessel which leaves port and steams directly to a point where other vessels are already reporting good schools of fish by radio). For this reason the use of effort and c.p.u.e. has been abandoned for some purse-seine fisheries. This is a policy of desperation unless some independent estimate of abundance is available (see section 2.3.6), and it is usually best to use the available effort data, though with caution in the interpretation of results.

Catchability and stock abundance can also be correlated without any causal relation existing. The common features of many fisheries is for catchability to increase with time, and for the stock to decrease under increasing fishery pressure. Thus, a falling stock is associated with increasing catchability, so that the observed c.p.u.e. underestimates the actual fall in stock abundance. The trend on catchability can therefore seriously affect the conclusions reached regarding the state of stock, and corrections have to be made for whatever improvements in vessel, gear, or fishery practices are causing the trend.

Correlations between the *amount of fishing* and catchability can be of either sign. Co-operation between vessels in locating good concentrations of fish can increase the catch of each vessel in a fleet compared with individual vessels acting entirely independently. Organized searching is a feature of some large fleets from socialist countries, but the individual fishermen of western Europe also depend very much on reports from other fishermen to know where to fish. On the other hand, on restricted fishing grounds, interference between boats and competition between them for the best places to shoot their gear can reduce the c.p.u.e. of each boat when many are present. A distinction should be made between this *interference*, in which the c.p.u.e. is reduced even if the stock abundance is unaltered, i.e. the catchability is reduced, and the effect of increased numbers of vessels on the stock abundance, i.e. the *effect of fishing*, in which the c.p.u.e. of all vessels, including those fishing the same stock later on other grounds, is reduced in proportion to the reduction on abundance.

Interference or co-operation will distort the linear relation that should ideally exist between the total fishing effort and the fishing mortality and corrections should, as far as possible, be made. Although quantitative estimates of their effects are not easily made by formal calculations, reasonably qualitative estimates can be obtained through familiarity with the fisheries and the fishermen—how much they complain about the interference of others, and how much they rely on information from others.

The biggest sources of variation in catchability are essentially *random*. A boat may take a good catch because it happens to steam across a good patch of fish, and detect it on its echo-sounder, or catch it in its trawl. Catches in a particular week can be particularly good or bad because the weather or other environmental conditions make the fish particularly available to the gear. For example, over most of the North Sea, east winds are associated with poor trawl catches. These variations are of prime interest to the fisherman since they determine his living, but on the scale of the individual trawl, or even for the fishery as a whole for a period of a few days, these variations are largely irrelevant to stock assessment studies. Large-scale variations, such as high catchability for most of a fishing season, can introduce undesirable variations into the analysis, which may be removed by analysing the effect of weather and similar factors (e.g. Gulland and Kesteven, 1964), but this is generally no more than a minor nuisance.

The effect can be more serious when policy decisions are made from information on catches in short periods which can be significantly influenced by environmental factors. For example, one effect of the El Nino that occurred in Peru in 1972 was that the stock was concentrated in a small strip near to the coast (Boerema and Gulland, 1973; Anon., 1973). This allowed good catches to be maintained even when the stock was rapidly declining; this in turn delayed the recognition by the Peruvian Government of the serious situation and the introduction of appropriate measures. Whenever attention is given to catches on c.p.u.e. in a particular short period, a check should be made on whether these catches or c.p.u.e. might be affected by short-term environmental factors.

2.3.3 The standardization of fishing effort

The fishing mortality generated (i.e. the proportion of the total stock removed) by a unit can be determined as the product of

(i) the size of the area influenced by the fishing gear in one unit of effort;
(ii) the proportion of the fish within this area that is retained by the gear;
(iii) the ratio of the density in the fished area to the average density in the whole area inhabited by the stock.

Viewed another way, the density of the stock will be equal to the catch taken by a unit of effort divided by the product of these three factors (area covered by unit effort, the proportion retained by the gear, and ratio of the densities in the fished area, and the whole area of the stock).

Therefore, the standardization of fishing effort and the corrections for any changes in catchability can in theory be done in terms of one or other of these factors. In practice the work has to be done in terms of the statistical data that are normally available in compiling fishing effort information; that is, records for each fishing unit of how much fishing it has done in rather simple terms. For these, fishing effort can be considered as the sum, over all units of the product of the fishing power of each unit, and its fishing time, or number of unit operations. These factors together are equivalent to the product of elements (i) and (ii) above.

Studies of fishing effort, and steps taken to standardize effort are therefore concerned with looking at the fishing power of individual units; measures of fishing time, or numbers of unit operations; and the geographical distribution of fishing relative to fish.

For some types of gear, which act by straining or selecting out the fish from a large volume of water (e.g. trawls, some types of seine), a first approximation to the fishing mortality can be given by the ratio of the volume (or area in the case of bottom-living fish) swept clear by the gear to the total volume (or area) inhabited by the stock. That is, the volume or area swept by the gear provides a direct measure of fishing effort (Treschev, 1975). For example, the fishing power of a midwater trawl, based on a fishing time of one hour, will be given by the product of the area of the mouth of the trawl times the distance travelled in one hour; in terms of area, the fishing power of a purse-seine, using a unit of one haul, will, if it approximates a circle, radius r, when shot, be equal to πr^2, or $\frac{1}{4\pi} \cdot l^2$, where l is the length of the net. In principle this approach can be used for a variety of gear, e.g. the power of a gill-net may be estimated as the area of the net times the average distance a fish might swim towards the net while it is in operation, or the power of a lobster pot is πr^2, where r is the average range within which lobsters can be attracted into the pot.

These latter estimates of swept area, and hence of fishing power, become somewhat speculative. Even for the more suitable gears the area or volume to be used may not be too clear. For example, for the normal bottom trawl with long bridles between the otter boards and the net, should the distance used be that between the boards (too large, because not all the fish are shepherded into the path of the net), or the actual width of the path of the net (too small)? A more serious difficulty is that the swept area estimate in itself takes little account of the proportion of the fish in the area that is not caught (which is probably highly variable between species and between different examples of the same basic gear), and no account of the fact that no gear is fished entirely at random and many are directed very carefully at high concentrations or individual schools. Although consideration of the swept area can be useful as a general guide to fishing power, and to possible changes in it due to changes in fishing practice, it cannot be recommended for obtaining quantitative estimates.

Fishing power can be measured quantitatively, relative to some standard fishing unit, by comparing their catches per unit effort, i.e. for the ith fishing unit, the relative fishing power P_i is given by

$$P_i = U_i/U_s \tag{2.10}$$

where U_i and U_s are the c.p.u.e.s of the ith and the standard gears respectively. If the interest in calculating fishing power is in studying those characteristics of the vessel or gear that determine catch rates, these c.p.u.e.s should be calculated on the same densities of fish, preferably at the same time and place. However, if the interest is solely in the effect on the stock, then it is only important that both gears are fishing on that stock, or on that area or other subdivision of the stock that is used in computing fishing intensity (i.e. an index of fishing mortality that takes some account of the distribution of fish and fishing—see later sections). Any ability of the fisherman operating a fishing unit to work on higher level densities than the standard unit improves its real fishing power and should be included in the analyses.

The simple expression in equation (2.10) allows the relative fishing power of a vessel to be estimated from data from those occasions when it is fishing in company with, or at least on the same stock as, the standard vessel. Because of the high variability associated with most fishing operations, estimates of fishing power from single comparisons are likely to be very variable. To combine data from several comparisons it is preferable, since catches usually have a log-normal distribution (i.e. the logarithm of the catches have a normal frequency distribution) to transform by logarithms, and write the pooled estimate of relative fishing power as $\bar{P}_i$. Where

$$\log \bar{P}_i = \frac{1}{n}\sum \log P_{ij}$$

$$= \frac{1}{n}\sum \log(U_{ij}/U_{s,j}) = \frac{1}{n}\sum (\log U_{ij} - \log U_{s,j}) \qquad (2.11)$$

where U_{ij}, $U_{s,j}$ are the catches per unit of the ith and the standard vessel on the jth of n-comparisons.

In practice, for each vessel there will only be a limited number of comparisons with the standard vessel, and estimates obtained from equation (2.11) will still be highly variable. The number of comparisons available can be increased by using a group of similar vessels as a standard, or by using one or more intermediate steps, e.g. comparing the ith vessel with vessel a, on some occasion j and the latter vessel with the standard, on another occasion, k, and writing

$$P_i = P_i/P_a \times P_a/P_s$$
$$= U_{ij}/U_{a,j} \times U_{a,k}/U_{s,k}$$

or, more conveniently, in logarithms

$$\log P_i = \log U_{ij} - \log U_{a,j} + \log U_{a,k} - \log U_{s,k}$$

This can readily be extended to a number of comparisons between each pair of vessels, and results in lengthy but basically simple computations. Robson (1966) proves a good description of the general approach, and Berude and Abramson give a computer program that is very suitable for handling the data.

The result will be some best estimate $\hat{P}_i$ of the fishing power of each vessel in the fleet. From these a figure for standardized effort could be obtained, as

$$f = \sum \hat{P}_i t_i \qquad (2.12)$$

where t_i is the fishing time of the ith vessel. However, this could be a lengthy process. Also, as new vessels join the fleet, estimates would have to be made of their individual fishing power. It is usually more convenient to find some easily recorded characteristic of the vessels or their gear (tonnage, horsepower, size of net, etc.) which is likely to be related to fishing power. Then the estimates of individual relative fishing power can be related to those characteristics, for example by fitting a linear regression,

$$P_i = a(GRT_i) + b$$

and equation (2.12) can be rewritten as

$$f = \sum (a(GRT_i) + b)t_i \tag{2.13}$$

or, if the relation between fishing power and some vessel characteristic, e.g. horsepower, can be treated as proportional, thus $P_i = cHP_i$
then

$$f = \sum cHP_i t_i$$

or, since we are concerned with relative units, the constant c can be dropped, and we can write

$$f = \sum HP_i t_i \tag{2.14}$$

The computation of total fishing effort can also be reduced, generally without significant loss of information, by considering groups or classes of vessels. The classes may be defined by a single characteristic—for example tonnage, using groups such as 30–49 tons, 50–69 tons, etc.—but additional divisions can be made to make the vessels within a group more homogeneous by dividing the 50–59 ton vessels into those built of wood and those built of steel, or those with engines greater or less than a certain horsepower.

The average fishing power of the vessels in each group can be estimated from the fishing power of the individual vessels. Alternatively, it can be estimated from the total statistics for the group, and some standard group, for some particular period, using equation (2.10) with the subscripts denoting classes rather than individual vessels.

Calculations of relative fishing power from records of commercial catches at a particular time of individual vessels, or groups of vessels, is satisfactory when the fishing power of most vessels in the fleet is the same from year to year, and the average fishing power changes because of the addition of new vessels, usually of greater efficiency, and in different tonnage or other class from the typical vessel of the existing fleet. Comparisons within a year cannot help detect and measure changes in the fishing power of a given vessel from one year to the next. These may involve subtle changes in the way the gear is rigged or used, which are difficult to deal with (though a scientist in close contact with the fishery should be aware that they are happening, and interpert the c.p.u.e. data with the possible effects of changes in mind). Sometimes more substantial changes occur, for example the adoption of heavy chains in front of the footrope by North Sea trawlers fishing for flatfish. In such cases the effect of the change can in principle be measured by controlled comparative fishing trials, usually with two ships fishing close together, of the pattern used when experimenting with fishing gear (cf. ICES, 1974). Data from research vessel surveys are also useful in dealing with the problem of learning during the early years of a commercial fishery (Brown *et al.*, 1976). This work is expensive in ship's time, but since the changes in the commercial fleet will often be the outcome of research carried out by gear technologists, the results of their studies and experiments may be available to make the calibration between old and new gears without the need for special work at sea. Year-to-year changes in the fishing power of individual vessels can be dealt with by repeating the studies for a series of years, provided some vessels do not change their characteristics from one year to the next. For example, Newman *et al.* (1978) carried out a study covering the South African purse-seine fishery in each year from 1964 to 1972, and determined a number of factors influencing fishing power, including the effect of changes of skipper. It is also possible, using methods such as those of Kimura (1980) to obtain estimates of relative fishing, and standardized measures of c.p.u.e. for a series of years from a single analysis.

2.3.4 Distribution of fishing

When the fishing power and fishing time have been fully standardized, the resulting catch per unit effort will be proportional to the average density at the positions fished, the average being weighted according to the amount of fishing at each place. This average density will almost certainly be greater than the true average density, because most fishing will be done on the grounds giving good catches. The c.p.u.e. will, however, still be a valid index of the density so long as the ratio of the true density to the density weighted by the amount of fishing remains constant. The difference between the true average density and the average density taken over the positions fished may be considered in two parts, corresponding to fishing tactics—getting the best catch on a particular ground—and fishing strategy—making the best choice of grounds, taking into account such things as different steaming time to different grounds, the different species composition on the grounds, and the different prices for different species, etc. On any given fishing ground, which for the North Sea trawl fisheries may be thought of as, say, 10 miles across, the distribution of fishing will be determined solely by the fishermen's ability to find the small local concentrations of fish. The density in the unfished parts of the fishing ground will be unknown. The fisherman will attempt to make the ratio of the density in the fished areas to the average density as high as possible. In the short run this ratio might be expected to be approximately constant, but over a longer period the introduction of new devices may permit the fishermen to concentrate more effectively on the fish, either directly (echo-sounding) or by more accurate navigation (e.g. Decca, radar, or echo-sounding). These improvements may be most easily considered as changes in the fishing power of the fishing unit (vessel, plus fishing gear plus other gear). In principle they can be analysed and corrections made, following the methods of section 2.3.3, e.g. comparing the c.p.u.e. of vessels with and without echo-sounder. However, this could give somewhat biased results, because it will usually be the better fishermen that first use a new type of equipment. If reliable estimates cannot be obtained by objective calculations, it may be possible to obtain some subjective estimates from talking to the fishermen. To the first approximation, though, we may write, for any one fishing ground

$$U = \frac{C}{f} = q'D' = qD \tag{2.15}$$

where $C =$ catch, $f =$ effort, q, q' are constants, $D' =$ average density, weighted by amount of fishing, and $D =$ true average density on that ground.

A unit stock will usually be distributed over several fishing grounds, and by expressing the total catch and effort as the sum of those in the individual grounds it can be seen that the ratio of total catch to total effort will be equal to the weighted mean of the c.p.u.e. in the various grounds, the weighting factors being the effort on each ground, i.e.

$$\frac{C}{f} = \frac{\sum C_i}{\sum f_i} = \frac{\sum\left(\frac{C_i}{f_i} \times f_i\right)}{\sum f_i} \tag{2.16}$$

where C, f are the total catch and effort, and C_i, f_i the catch and effort in a particular fishing ground.

Since the distribution of fishing is likely to vary from year to year due to changes in the relative abundance of different stocks, etc. these weighting factors will vary, and the ratio of the total catch per unit effort to stock abundance will also vary (unless the density of fish on all grounds is the same).

However, from equation (2.15) we can for any fishing ground, i, whose area is A_i say, express the number of fish as

$$N_i = A_i D_i = \frac{A_i}{q_i} \times \frac{C_i}{f_i}$$

If the whole range of the stock can be subdivided into regions, within each of which equation (2.15) can be applied, then by adding, the total number in the stock is

$$N = \sum_i N_i = \sum\left(\frac{A_i}{q_i} \times \frac{C_i}{f_i}\right)$$

If q_i is constant $= q$ for all grounds, then

$$N = \frac{1}{q}\sum\left[A_i \times \frac{C_i}{f_i}\right]$$

and the overall density is

$$= D = \frac{N}{A} = \frac{1}{q} \times \frac{1}{A} \times \sum\left[A_i \times \frac{C_i}{f_i}\right] \tag{2.17}$$

where $A = \Sigma A_i$, i.e. the density is the weighted mean of the catches per unit effort in each subregion, the weighting factors being the areas of the regions. If all the areas are of equal size, this reduces to

$$D = \frac{1}{nq} \sum \frac{C_i}{f_i}$$

where n = number of areas.

The effective total effort (i.e. the measure of effort which will remain proportional to the fishing mortality regardless of changes in the distribution of fish and fishing) can be calculated by dividing the index of density derived from equation (2.17) into the total catch, i.e.

$$f = \frac{C}{qD} = \frac{A \sum C_i}{\sum \frac{A_i C_i}{f_i}} \tag{2.18}$$

The effective effort per unit area (f/A), obtained from this equation can be referred to as fishing intensity.

These formulae also allow density indices to be obtained for any particular subgroup of the population, for example an age group. If the density index for the entire population is given by total catch divided by total effort, then the index for a given age group is found by dividing the number landed of that age group by the total effort. Otherwise the index is obtained by raising the number of each age in a unit weight by the weight caught per unit effort. If there are marked differences in composition between different regions of the stock then the density indices will have to be obtained for each region separately, and the index for the whole stock obtained by weighting up by the areas of each region. The size of these regions should be small enough to ensure uniform composition within them, but in general should be larger than the separate grounds which, as discussed earlier, are used in giving the overall density index.

The problem in using this method to deal with possible changes in the distribution of the fishery relative to that of the fish is that it requires information from each subarea into which the distribution of the fish stock is divided. This requirement can be relaxed by making assumptions or interpolation in respect of the subareas in which no fishing was done, but if the usefulness of the method in reducing the effect of change in the distribution of fishing is not to be lost, information must be available for at least the majority of subareas. The choice of the size of subareas then runs into difficulties. On the one hand if they are made small and numerous there will be too many blank areas in which there is no fishing. On the other, if they are large, there can be significant changes in the distribution within a sub-area which will change the relation of c.p.u.e. to density.

Another aspect of the same problem—failure to obtain data from all areas—arises from the fact that fishermen will only stay at places where catches are at least moderate. If they do visit areas where the average density is low, they may stay there only long enough to contribute to the catch and effort statistics if the density when they fish is exceptionally high for that subarea. Thus, the commercial c.p.u.e. data are likely to give a biased impression of the relative density in the less productive areas. This problem is most serious when the fishery is based on a single species. It is less serious when the fishery is directed at a variety of species with different distributions. Then the density of all species together may be high enough to attract commercial fishing in subareas in which the density of individual species may be low. The method is therefore of particular value in multi-species fisheries such as the North Sea trawl fishery (in which data are routinely collected in sub-areas of approximately 30 n.m square), or the tuna long-line fishery (in which sub-areas of 5° square are used). The practical application of these data is much simplified by using computers, not only for calculations but also for graphical presentation (Loh Lee Low, 1974).

2.3.5 Multi-species problems

Every fishing gear can catch a large variety of species, and many different species occur on most fishing grounds. Very few fisheries are based solely on a single species. In practice the interpretation of catch and effort data concerning one species has to take into account the effect on the fisherman's tactics and strategy of possible catches of other species. At the extremes there are two relatively simple situations. In the one case as in the Peruvian anchoveta fishery during, the 1960s one species is so important that the fishermen's policies are in no way affected by the abundance of other species. In the other case, the species concerned is so scarce that, though valuable enough to be kept when caught, its abundance is never so significant as to affect policy—most of the individual species in the multi-species tropical trawl fisheries provide examples of the latter. In the intermediate cases some allowance has to be made in any interpretation for the possible effects of changes in species preference on the distribution or methods of fishing and hence on the catchability coefficient for individual species or methods.

A method for dealing with large-scale changes in distribution is described in section 2.3.4, using equations (2.16) and (2.17). A method of dealing with other changes is to collect information on the main species that is the objective of each fishing operation (trip, day fishing, etc.) as part of the standard collection of statistical data. Then the estimate of c.p.u.e. for each species can be based only on those operations for which it was the target species. Ideally, this preference should be expressed as an *intention* before fishing starts rather than a reflection of the success of this intention after the event. Otherwise the resultant catchability coefficient will be unsatisfactory, since it will tend to increase at low real densities of the species concerned because only unusually productive operations, i.e. those actually catching good quantities of the species, will be included in the calculations. This is perhaps asking a lot of the fishermen to record intention properly so as to include the occasions when he would have liked to catch, say, yellowfin tuna if he could, and had to make do with, say, skipjack.

A similar method, which depends less on direct information from the fishermen is to include in the analyses only those landings that contain more than some threshold percentage of the species concerned (Ketchen, 1964). This suffers from the same disadvantage that the catchability coefficient will tend to increase as the stock decreases through the selective inclusion of only the more successful trips.

Interpretation of catch and effort data for one species can be affected by other species through some of the other mechanisms already discussed. For example, in most types of saturation the effectiveness of a unit of effort is reduced by the presence of any species, and not merely the species of direct interest. The fishing power of a tuna long-line is reduced as much if the bait is taken by a shark as if it were taken by a tuna. Again, in a fishery where searching is important and the amount of effective searching done per day is affected by the loss of time in setting and hauling the net and handling the catch, the abundance of all species of interest affects this 'lost' time and hence the effective effort. The question of searching time in a multi-species fishery has received particular attention in relation to the tuna purse-seine fishery in the eastern Pacific, in which skipjack is an important alternative to the preferred species—yellowfin tuna (Pella, 1969).

2.3.6 Problems of particular fisheries and gears

Each of the aspects of definition and measurement of fishing effort discussed in the previous sections occurs in any fishery and in relation to any fishing gear, but for each gear and fishery there are certain aspects that are particularly important. These are discussed below in relation to the major types of gear and summarized in Table 2.1 (from ACMRR, 1976). A similar examination directed specifically at shellfish fisheries is given by Caddy (1979).

(*i*) *Danish seine*

The fishing power of a Danish seine depends mainly on the size of the gear which, in most Danish seine fisheries, is related to the size of the fishing vessel. Therefore, classification of effort measures by vessel size categories will largely take account of this effect of gear size. The time unit for the Danish seine can be either the time the net is actually fishing or the number of sets. But owing to difficulties in determining the former in each haul (the time between shooting and hauling the gear may be markedly different for consecutive hauls made at different depths and yet the actual fishing time may be the same), the latter is the more appropriate measure. Effort in the Danish seine fishery should therefore be recorded as number of sets by vessel category.

(*ii*) *Bottom trawl*

The fishing power of a bottom trawl depends on its size and speed through the water. These are usually closely related with the size and/or engine power of the vessels. Many studies have shown a linear relationship between size or engine power of the vessel and its catch rate. Therefore, the collection of effort data by vessel size or engine power category may eliminate the need for information on gear size and other characteristics correlated with size or power category. Different types of bottom trawl (e.g. single boat otter side trawl, stern trawl, pair trawl, beam trawl, single or double rig shrimp trawl, etc.), distinguished in international gear classifications, have different fishing powers and should be kept separate in the data collection.

In several bottom-trawl fisheries searching is of minor importance. In such fisheries the total time of actual fishing with the gear on the bottom (towing time) gives the best effort estimate, although the number of days fishing will give a reasonable approximation. Other approximations, such as total number of days on the grounds, or days absent from port, are useful if more detailed data are unavailable. In such cases it would be necessary to make corrections for time required for steaming, or for loss of fishing time due to weather conditions, if the proportion that these are of the time as recorded change from year to year.

If, however, searching plays a more substantial role, total number of days spent fishing is likely to give a more appropriate measure than total towing time. If possible, information on actual fishing time and time spent searching should be collected to enable detailed studies to be made of changes in the fishing operations and of the importance of searching as a component of the effort unit.

(iii) Midwater trawl

In midwater trawling, engine power is the key factor determining the size of net used, and the speed of towing, and hence the fishing power of the vessel. Fishing effort data should therefore be collected by vessel engine power categories. Since in almost all midwater trawl fisheries, searching (with echo-sounders or sonar) forms an important component of the individual fishing operations, the time actually spent fishing is an inadequate measure of effort. Its use alone may give rise to serious bias. Number of days fishing is likely to be a more satisfactory measure for midwater trawl fisheries, but it is also very desirable to collect detailed data on searching time and actual fishing time (if necessary on a sampling basis) so that comparison can be made of the various estimates and eventual later correction of the catch per day fishing data. Searching time can eventually be determined from information on the number of days fishing, and the actual towing time.

(iv) Encircling gears

The encircling gears such as purse-seine and ring-net are usually directed towards schooling fish species and they must be used by aimed fishing. Catch per unit effort is therefore strongly influenced by the skill of the fishermen (in detecting the fish school and in gear handling) and the behaviour of the fish as well as by the technological characteristics of the fishing unit.

Among the technological characteristics the size of vessel, its gear and its engine power may all be equally important in particular fisheries. For example, the power of the engine may influence the speed and hence success of a set within a vessel size category. It may therefore be necessary to combine several parameters to arrive at an appropriate fishing power index.

Many encircling fisheries are based on low-value, large-quantity species, e.g. anchovy or sardine, and are particularly susceptible to saturation of the handling and carrying capacity of the vessel so that the unit of time chosen in calculating effort should be small enough to exclude this effect, i.e., c.p.u.e. should be calculated as catch per shot, or catch per hour on the fishing grounds, rather than catch per day on the fishing ground or per trip. Alternatively, provision must be made to identify and eliminate from the analysis periods when saturation rather then fish abundance has limited the catch.

The reliability of c.p.u.e. as an index of changes in abundance in these fisheries is very sensitive to the way in which the distribution, shoaling characteristics, and density change with changes in total abundance. Decreased abundance may be reflected in one or a combination of (i) smaller average school size, (ii) a greater distance between schools, or (iii) a reduction of the total area inhabited by the exploited stock.

In most cases the abundance of fish is best measured as catches taken, or schools seen, per unit search time. Statistics of fishing time should therefore be recorded in such a way as to make the actual searching time easily separable from other time at sea. The possibility of changes in the extent of the area inhabited by the stock also requires examination. In a purse-seine fishery, more than any other fishery, the choice of effort units for estimating c.p.u.e. must be made very carefully, and the data interpreted with great caution and, wherever possible, corroborated with independent sources of information. Alternative methods of obtaining indices of abundance are discussed by Ulltang (1977).

(*v*) *Gill-net*

The most important power factor is the size of gear, which may be measured as number of nets since they are often of a standard size, though this should not be assumed always to be so without checking. Number of nets x days fished will probably give an appropriate measure of effort in most fisheries. Care should, however, be taken to watch for significant changes in the fishing gear, e.g. in size of nets, the materials in which they are made, the way they are rigged, or the average duration for which they are fished. If they do change, the possible effect on fishing effort should be checked, and adjustments made if necessary to the measures used. Searching time may be an important factor in drift gill-net fisheries aimed for schooling species, but not in bottom gill-net fisheries.

(*vi*) *Hooks and lines*

In most bottom and pelagic long-line fisheries the fishing power in hook-and-line fishing is mostly governed by the number of hooks operated during a fishing operation, though for convenience they are often recorded in units (e.g. baskets or 'skates') which usually have a standard length and a standard number of hooks. In some other kinds of line fisheries, on the other hand, only a small number of hooks can be efficiently operated simultaneously from the same boat. For example, in the albacore troll fishery in the Bay of Biscay, boats are all rigged in the same way with the same number of lines. In the latter fisheries, therefore, fishing power may be expressed simply as the number of fishing units in operation, where necessary divided into vessel (or crew) size categories. It is necessary in using such a measure to make detailed periodic surveys to check whether significant changes have taken place in boat/gear/crew characteristics, in gear design and rigging, or in fishing practices. Changes in rigging may be particularly important in affecting catching efficiency. For example, Skud (1972) has shown that an increase in the number of hooks per unit length increased the catch of small Pacific halibut, but not that of larger halibut. Skud (1978) discusses this question further, including the effect on species other than halibut.

The type of bait—natural and artificial—can affect the catchability in most fisheries, and information on bait has to be regularly collected. In pole-and-line fishing, limited bait supply may significantly reduce the amount of the total time at sea a boat can devote to tuna fishing. The resulting possible bias would be obviated if time spent for searching and capturing tuna is recorded separately from the time devoted to bait fishing and other non-directly productive time components.

Gear saturation effects are likely to occur in almost all kinds of line fishing; relevant information should therefore be regularly collected on these parameters and incorporated in the calculation of effort and abundance indices.

In most bottom-line fisheries, acoustic instruments are mainly used for better location of fishing grounds (through observations of bottom topography), and seldom for the direct detection of fish concentrations. Fisheries with lines for small pelagic fish species are mostly of a small-scale nature, in which the use of electronic equipment is limited. In pelagic tuna fishing, school detection is mainly made through the observation of associated fauna (flocks of birds, porpoises, etc.) and the impact of acoustic equipment on fishing efficiency is therefore likely to remain limited.

Active searching is important, and should therefore be recorded as such, in most surface tuna fisheries (e.g. pole-and-line), whereas its importance in hand-lining for groundfish would usually be negligible. Pole-and-line represents an extreme case in which trip duration has to be properly allocated according to its four major components: steaming between port and fishing grounds and other dead times, fishing for bait, searching for suitable tuna schools, and finally catching.

In some small-scale inshore line fisheries, fishing operations and trips are often of constant duration, for example when they are adjusted to the tidal cycle. Similarly, in tuna long-lining soaking time remains more or less constant for each set. Measurement of fishing time can then be made by simply recording the number of sets, or in some cases the number of trips. But, again, periodic checks should be made of changes in fishing practices which affect fishing efficiency.

Based on the above factors, fishing effort for the various types of line fishing would be best measured in the following units:

—hand-lines/groundfish	Number of lines, soaking time (or trips)
—long-lines/groundfish, tuna	Number of hooks, number of sets
—troll/tuna, salmon, small pelagic	Time on fishing grounds, adjusted for number of lines
—jigging/squid	Number of nights, adjusted for number of gear units
—pole-and-line/tuna	Searching time and time on tuna fishing grounds, adjusted for number of hooks with, according to the particular fishery, additional adjustments for differences in bait, rigging, searching, number of gear units, etc.

(*vii*) *Light fishing*

The use of light for attracting fish, for example in some purse seine, lift-net, and jigging fisheries, brings a particular set of factors to the measurement of c.p.u.e. In such fisheries, the fishing effort is directly related to the attractive power of the light source (intensity, wavelength, immersed or not, etc.), the characteristics of the catching gear, and to the number of nights fished. As the attractive efficiency of artificial light is affected by moonlight, the phase of the moon—and possibly seasonal changes in cloud coverage—can affect catches, and will have to be accounted for if the pattern of fishing in relation to moon-phase may alter.

Sometimes searching is carried out during daytime before initiating attraction of fish (e.g. in purse-seining with light). Effects of searching on the fishing effort estimates could often be accounted for by refined time/space resolution in data collection; otherwise catches would have to be referred to each 24-hour period each vessel spends at sea.

(*viii*) *Lift-nets*

The main sources of bias in lift-net fisheries are to be expected from incorrect estimation of fishing time and from improper allocation between searching time and time actually spent fishing. The fishing power is relatively less affected by vessel characteristics, though account should be taken of the number of gears which can be simultaneously operated from a single boat (i.e. to express c.p.u.e. as catch per net, not catch per boat). Searching efficiency can be significant in determining fishing power.

(*ix*) *Pots and traps*

The main fishing power component in pot fisheries is the number of pots set, and effort should be recorded accordingly. Boat characteristics do not directly affect fishing power, but only indirectly through the number of pots carried per trip. Records should be kept of the design of the pots or traps, the kinds of bait used, and the length of time the gear is left on the bottom. Adjustments will need to be made to the reported effort data if these change. The effective fishing effort can be expected to increase as the duration between lifts increases, but less than proportionally. That is, if, for example, the interval between lifting lobster pots is changed from every day to every second day, the catch per lift will go up, but the catch per lift-day will go down. In such cases studies will need to be made of the quantitative relation between the catch rate (i.e. the effective fishing power), and the length between lifts. For example, Munro (1974) has studied the operations of fish traps in the Caribbean. He showed that the number of fish present in the trap, which is a balance between those entering and those escaping, increases for the first few days, and then tends towards an asymptote, or even decreases.

(*x*) *Fixed traps*

These are probably the fishing gears for which sources of bias are the smallest—at least for a given unit of gear no influence of operating crew or boat is to be expected. However, changes in catching capacity are to be expected according to its size and possibly also from the way the net is mounted and set. However, these gears and some of the passive ones (pots, gill-nets, lines, etc.) do depend on the appropriate behaviour of the animals. This behaviour may be affected by changes in density, e.g. due to crowding, or in the local environment. For example, coastal traps may be less effective if the waters become polluted.

(*xi*) *Other gears*

A variety of other gears are used in small-scale inshore fisheries, especially in tropical areas. On the whole the ways of operating these traditional gears change little, so that no correction is necessary for improvements in fishing power. The simplest measures, such as numbers of fishermen or fishing units, recorded perhaps once a year, can usually therefore be used as measures of fishing effort, though the scientist should, as always, keep in close contact with the fishery to check that no changes in fishing practice are in fact occurring.

2.3.7 General procedures

From the preceding sections it is clear that there is no general set of rules that can be directly followed in a simple manner in order to derive the desired indices of fishing mortality and abundance from catch and effort statistics without some careful thought. Some general procedures can, however, be suggested which will help these thoughts to be focused on the right subjects at the right time, and hence help produce the best indices from the available data.

The first rule is that in any fishery for which data are available from different sources or presented in different ways, all data should be examined, and also kept separate in the early analysis. For example, if the same stock is being harvested by different countries or even from different ports, or by different types of gear, or by obviously different size classes of vessel, the data from each source should be kept separate. Each set of data should then be examined for possible trends in catchability (e.g. increases in vessel size) and adjustments made along the lines described in section 2.3.3. This examination is best done by talking with fishermen and others who have been familiar with the fishery over the period and finding out from them what changes and developments have occurred. Where relevant indices of c.p.u.e. and effort should also be obtained which have been corrected for possible changes in the seasonal or geographical pattern of fishing, using the methods of sections 2.3.1 (equation 2.4) or 2.3.5 (equations 2.16 and 2.17).

Under ideal conditions these corrections should result in the data set from each source providing equally valid indices of abundance and fishing mortality. The next step is to check the extent to which the hopes that these conditions are met in reality have been satisfied by comparing the different indices. The comparison can be done in terms of either c.p.u.e./abundance or effort/mortality, though it may be easier to understand the causes of differences when considering c.p.u.e./abundance. For ease of comparison (e.g. between the catch per hours fishing of large trawlers and the catch per set of bottom gill-nets) it is desirable to express the c.p.u.e. of each gear in each year as a percentage of the c.p.u.e. of that gear in some standard year, or period of years. Any other index of abundance that might be available, for example from research vessel surveys of one type or another (see section 2.4), should also be examined and expressed in the same way. These indices should then be presented graphically, preferably on a logarithmic scale which permits an easier comparison between indices of abundance expressed in quite different units.

If all the c.p.u.e. data (and other indices if available) are in fact measuring changes in the abundance in a satisfactory manner, then all the lines in the figure should coincide with the actual changes in abundance, as a percentage of the average in some basic year(s). Some differences may occur because different fisheries may not be sampling the same part of the stock. Fisheries on older fish may be expected to show fluctuations due to natural fluctuations in the stock (e.g. in the strength of year-classes) which are out of phase with those shown by fisheries on younger fish, the difference in timing being equal to differences in mean age. Also, the abundance of older fish will be more seriously affected by fishing, so that as fishing increases the c.p.u.e. of fisheries on these age-groups may be expected to decline more rapidly than others. Discounting these differences and small random fluctuations, any deviations between the lines of c.p.u.e. from different sources indicate that at least one of the data sets is not providing a satisfactory index of abundance.

The most likely cause is that in one set there has been an increase in the effectiveness of a unit of effort through the introduction of new or improved fishing methods that have not been taken fully into account. Thus, the line trending upwards (or decreasing more slowly) should be treated with the greatest suspicion, though both (or all) sets of data should be examined for possible uncorrected changed in catchability. This examination may suggest corrections, e.g. for the introduction of improved acoustic gear in a purse-seine fishery, which bring the sets into conformity. Alternatively, it may suggest that one or more sets of c.p.u.e. is by its nature unsuitable for estimating abundance; these data should therefore be rejected for further analysis. At the worst no set of c.p.u.e. data may seem suitable—as was found in the case of the Norwegian mackerel data—and other methods will have to be used to monitor abundance changes (see section 2.4).

More usually, some data sets can be accepted as usable. If there is just one set, then the effort and c.p.u.e. of the fleet concerned can be taken as standard, and the total effort written as

$$\text{total effort} = \frac{\text{total catch}}{\text{c.p.u.e. of standard fleet}} \tag{2.19}$$

It should be stressed that if the data of the standard fleet can be expected to satisfy the basic condition that the appropriate catchability coefficient q is constant and information is sufficiently extensive so that random variations are not likely to be large (e.g. the figures will not be significantly changed by one single lucky trip), then the procedure of equation (2.19) will be more satisfactory than attempting to calibrate the effort of each sector of the fishery in some standard units.

If more than one set of data are acceptable, then they can be combined by expressing all the effort data in terms of one or other set as standard (the choice is immaterial), using equation (2.10) from which a pooled c.p.u.e. can be derived to use in equation (2.19).

Direct calibration of the effort data is possible by the use of cohort analysis (see section 4.3.4). This provides estimates of fishing mortality to be obtained for each year, and also each age-group. Dividing these by the nominal effort enables estimates of q to be obtained for each year (and if wished, also for each age). These can then be examined to determine if there is any trend with time or relation with stock abundance. This may reveal that catchability increases with decreasing stock size even for gears (e.g. trawls, Houghton and Flatman, 1980) for which this effect is not often suspected. The advantage of this approach is that by obtaining a better insight into possible changes in q, it enables effort and c.p.u.e. data to be used for the most recent years for which cohort analysis is less reliable.

2.4 SURVEYS

2.4.1 Methods and objectives of surveys

Apart from the commercial fishery, the other main sources of data in stock assessment are surveys carried out by research or similar vessels. The details of how surveys should be carried out, and the data from them collected and analysed are described in a number of FAO manuals (Alverson, 1971; Mackett, 1973; Gulland, 1975; Saville, 1977; Burczynski, 1979), and will not be repeated here. For the present it is only important to note what types of information can be provided from surveys that will be useful in stock assessment, and to outline briefly the advantages and disadvantages of the different methods of surveying by which this information can be collected.

Survey data can be used in stock assessment in two main ways: first, for monitoring, that is, to provide at regular intervals (most conveniently annually) indices of stock abundance; second, to produce estimates of absolute abundance, possibly at only one instant of time, and most usually in advance of intense exploitation.

As was noted in section 2.3, c.p.u.e. data from some part of the commercial fishery usually provides the most convenient index of stock abundance, but for some stocks there may be no c.p.u.e. data that is satisfactory. This may be because, over a wide range of stock sizes, the observed c.p.u.e. is only weakly related to stock size (e.g. some purse-seine fisheries on pelagic fish), or because there have been substantial but non-quantifiable changes in the fishing power of the vessels, or the strategy of the fishermen (e.g. changes in species preference in some trawl fisheries). A monitoring survey repeated at regular intervals, in which the methods used are maintained constant from year to year, will provide an index of abundance that is free of the difficulties caused by possible changes in the catchability coefficient q; this index can be used as in equation (2.19), to divide into the total catch to provide a measure of total effort. The similarity in the uses in later assessment analyses of commercial c.p.u.e. data and of survey data is most obvious when the survey is carried out with standard commercial-type fishing gear, and the abundance index can be expressed as, for example, catch per haul, but other output from surveys of, for example the integrated strength of signals received during an echo-survey can be used in exactly the same way.

Monitoring surveys of this type present few problems relating to stock assessment theory; provided the gear used can provide observations of the stock of interest, these can be used to give indices of abundance and hence of fishing mortality. The main question relates to the population being sampled, which may not be the same as the group of fish (normally only those of a size big enough to be caught by commercial gear) treated by most assessment analysis. This can be an advantage in that surveys with non-commercial gear can provide information on groups of fish not otherwise sampled—for example, surveys with small-meshed trawls can provide indices of abundance of young, pre-recruit year-classes. There can be difficulties when the survey information does not distinguish different stocks. Acoustic surveys are particularly subject to this disadvantage, and the actual acoustic work usually has to be supplemented with fishing (by the survey vessel itself, or another) to identify which species of fish is providing the echoes.

The main problems with monitoring surveys are operational ones of balancing cost against the value of the information obtained, and of determining the sampling design and survey pattern which will provide the best information at least cost. Monitoring surveys are expensive—usually requiring several weeks work per year of a costly vessel—but the cost does not rise in proportion to the size and importance of the fishery. These are therefore most likely to be worth doing in respect of large and valuable fisheries—as they are for example in the Thai trawl fishery, or on Georges Bank in the north-west Atlantic. They may be less easily justified, and the decision to engage in them should be taken with more care, in respect of smaller fisheries, for which the cost of regular monitoring surveys could be excessive, and jeopardize the chances of support for other important research.

Surveys that can produce absolute estimates of stock abundance introduce a new type of information into assessment work. The ability to use these estimates, in combination with data of total catch, to provide estimates of fishing mortality in absolute terms clearly makes much of the analysis of mortality rates (see Chapter 4) much simpler. In addition, estimates of total stock abundance, combined with estimates of natural mortality or other measures of turnover rate, can provide first approximations to the potential yield from the stock (see section 2.4.2 below).

Apart from the operational questions of cost and survey design, common to monitoring surveys, surveys aimed at producing estimates of absolute abundance have to face questions of possible bias. For example, the data from trawl surveys have frequently been used to provide estimates of absolute density on the basis of the catches representing some fixed percentage (often 50 per cent) of the actual stock occurring in the area covered by the trawl. Any departure of the real percentage from 50 per cent will bias the results, and the available information (e.g. Edwards, 1968) suggests that the true percentage varies greatly between species. Other types of survey are subject to similar biases. For example, surveys with precision acoustic instruments can measure the returned signals accurately, but depend on knowing the target strength for each species—which may not be known well, and in addition several species may by involved. Egg surveys also may involve difficulties in identifying species, and to use either egg or larval surveys in estimating abundance assumptions have to be made about spawning frequency and mortality and development rates of eggs and larvae. In addition, like monitoring surveys, surveys for absolute abundance are expensive and, unless carefully planned and carried out, can be subject to considerable sampling error. Therefore it cannot be recommended or expected that these surveys should always be carried out. However, the possibilities of surveys should always be considered. In addition, the opportunities of obtaining survey information for ships not engaged in direct fishery research (e.g. the examination of general plankton samples for fish eggs, or obtaining regular echosounder records from commercial vessels) should not be neglected.

2.4.2 Estimating sustainable yield from surveys

The data from surveys will usually be used together with data from other sources to carry out assessments using the techniques described in later sections. Survey data can also be used more directly to make assessments. Several types of survey give—with the reservations noted in section 2.4.1—estimates of total biomass. This estimate is interesting, but seldom exactly what the fishery administrator or planner wants to know; he usually needs to know how much can be caught each year. This quantity is clearly related to the biomass, or standing stock; other things being equal, the bigger the biomass the bigger the sustainable yield. Further, the ratio of sustainable yield to biomass must be connected with the turnover rate (growth and mortality rates) of the species concerned. For a given biomass the sustainable yield from a long-lived species will be less than that from a short-lived species.

This suggests that, for surveys of unexploited stock, the sustainable yield may be estimated by an expression of the form

$$Y_{max} = aMB_{\infty} \tag{2.20}$$

where B_{∞} = unexploited biomass, and M = natural mortality. Theoretical considerations suggest that the value of a is likely to be around 0.5 or somewhat less, so that a convenient expression for the sustainable yield is

$$Y_{max} = 0.5\,MB_{\infty} \tag{2.21}$$

Practical applications of this formula have shown that in general it gives useful results. It is obviously approximate, and should not be considered as a substitute for more detailed assessments. At the same time it is one of the few methods that can be readily used before fishing begins, and in particular at the moment when plans are being drawn up to start exploitation of a stock. At this time a rough estimate (accurate to within say 50 per cent) is all that is required.

Apart from estimates of biomass, application of this method requires estimates of M. If the biomass is obtained by trawl or other fishing surveys, then samples from the catch (especially if they can be aged) can be used. Otherwise rough estimates of natural mortality can be obtained by comparison with known values for similar species (see section 4.3.7). These estimates will inevitably be rough, but in most cases sufficient.

Surveys will not always be of wholly unexploited stocks. For example, there may be a small inshore artisanal fishery, and a survey may be carried out to determine the possibility of expanding the fishery offshore with larger vessels. For such surveys, equation (2.21) is clearly not appropriate. Compared with an unexploited stock the biomass will have been reduced by fishing, while the total mortality has been increased. This suggests that a suitable modified formula would be

$$Y_{max} = 0.5ZB \tag{2.22}$$

where Z = total mortality coefficient.

This is convenient if the total mortality can be estimated. For some stocks though, the best estimate of mortality may still be that of natural mortality secured from comparison with other species or stocks. For these, a better form is obtained by noting that $ZB = (F + M)B$ and that the catch $Y = FB$.

Therefore, we can write

$$Y_{max} = 0.5(Y + MB) \tag{2.23}$$

In view of all the economic and social uncertainties in starting up a new fishery, let alone the biological ones, realistic plans for the initial development will seldom aim to catch more than a fraction of the estimated sustainable yield. As these plans are put into effect, and effort increases, then there will be opportunities to make assessments by other, more precise methods. [Recent studies suggest that putting $a = 0.5$ gives too high values of potential yield and a more conservative value around 0.3 would be better.]

REFERENCES AND READING LIST

2.1 The unit stock

Cushing, D. H. (1968) *Fisheries Biology. A Study in Population Dynamics.* Madison, University of Wisconsin Press, 200 pp.

Cushing, D. H. (1975) *Marine Ecology and Fisheries.* Cambridge, Cambridge University Press, 278 pp.

Cushing, J. E. (1964) The blood groups of marine animals. *Adv. Mar. Biol.*, **2**: 85–121.

Gulland, J. A. (1980) Some problems of the management of shared stocks. *FAO Fish. Tech. Pap.*, 206.

Harden-Jones, F. R. (1968) *Fish Migration.* London, Edward Arnold, 325 pp.

Ricker, W. E. (1972) Hereditary and environmental factors affecting certain salmonid populations. In *The Stock Concept in Pacific Salmon* R. C. Simon and P. A. Larkin (eds.). H. R. MacMillan lectures in fisheries. Vancouver, BC, University of British Columbia, pp. 19–160.

Schaefer, M. B. (1957) A study of the dynamics of the fishery for yellowfin tuna in the eastern tropical Pacific Ocean. *Bull. I-ATTC*,**2** (6): 247–285.

2.2 Data from the commercial fishery

Anon, (1981) Methods of collecting and analyzing size and age data for fish stock assessment. *FAO Fish. Cric.* 736, 100 pp.

Bazigos, G. P. (1974) The design of fisheries statistical surveys—inland waters. *FAO Fish. Tech. Pap.*, No. 133: 122 pp.

Brander, K. (1975) Guidelines for the collection and compilation of fishery statistics. *FAO Fish. Tech. Pap.*, No. 148: 46 pp.
Deming, W. E. (1950) *Some Theory of Sampling.* New York, Wiley, 602 pp.
Gulland, J. A. (1966) Manual of sampling and statistical methods for fisheries biology. Part 1. Sampling methods. *FAO Man. Fish. Sci.*, (3): 87 pp.
Holden, M. J. and D. F. S. Raitt (1974) Manual of fisheries science. Part 2. Methods of resource investigation and their application. *FAO Fish. Tech. Pap.*, No. 115, Rev. 1: 214 pp.
Henderson, H. F., R. A. Ryder and A. W. Kundhongania (1973) Assessing fishery potentials of lakes and reservoirs. *J. Fish. Res. Board Can.*, **30** (12), Pt. 2: 2000–2009.
Parrish, B. B. (ed.) (1962) Requirements and improvements of fishery statistics in the North Atlantic region. Report based on documents of the Expert Meeting on fishery statistics in the North Atlantic area. Edinburgh, Scotland, 22–29 September 1959. *FAO Fish. Rep.*, No. 3: 217 pp.
Ryder, R. A. (1965) A method for estimating the potential fish production of north temperate lakes. *Trans. Am. Fish. Soc.*, **94**: 214–218.
Williams, T. (1977) The raw material of population dynamics. In *Fish Population Dynamics*, J. A. Gulland (ed.). London, Wiley, pp. 27–45.
Yates, F. (1953) *Sampling Methods for Censuses and Surveys.* London, Griffin, 401 pp.

2.3 Effort and catch-per-unit effort

Anon., (1973) Tercera sesión del panel de expertos sobre la dinamica de la población de anchoveta Peruana. *Bol. Inst. del Mar del Peru*, **2** (9): 525–599.
ACMRR (1976) Monitoring of fish stock abundance. The use of catch and effort data. A report of the ACMRR Working Party on fishing effort and the monitoring of fish stock abundance. Rome, Italy, 16–20 December 1975. *FAO Fish. Tech. Pap.*, No. 155: 101 pp.
Beverton, R. J. H. and S. J. Holt (1957) On the dynamics of exploited fish populations. *Fish. Invest. Minist. Agric. Fish. Food U.K.* (*Series* 2), No. 19: 533 pp.
Boerema, L. K. and J. A. Gulland (1973) Stock assessment of the Peruvian anchovy and management of the fishery. *J. Fish. Res. Board Can.*, **30** (12), Pt. 2: 2226–2235.
Caddy, J. F. (1979) Some considerations underlying definitions of catchability and fishing effort in shellfish fisheries, and their relevance for stock assessment purposes. *Manuscr. Rep. Ser. Mar. Sci. Dir. Can.*, No. 1489: 19 pp.
Caveriviére, A. (1978) Standardisation des effort des pêches des chalutiers ivoriens et estimation de l'abundance relative dans les divers secteurs. *Doc. Sci. Cent. Rect. Oceanogr. Abidjan ORSTOM*, 9 (1): 51–72.
Clark, C. W. and M. Mangel (1979) Aggregation and fishery dynamics: a theoretical study of schooling and the purse seine tuna fisheries. *Fishery Bulletin* **77** (2): 317–337.
Gulland, J. A. (ed.) (1954) Symposium on the measurement of the abundance of fish stocks. *Rapp. P.-V. Reun. CIEM*, 155: 223 pp.
Gulland, J. A. (ed.) (1955) Estimation of growth and mortality in commercial fish populations. *Fish. Invest. Minist. Agric. Fish Food U.K.* (*Series* 2), **18** (9): 46 pp.
Gulland, J. A. (ed.) (1956) On the fishing effort in English demersal fisheries. *Fish. Invest. Minist. Agric. Fish. Food U.K.* (*Series* 2), **20** (5): 41 pp.
Gulland, J. A. and G. L. Kesteven (1964) The effect of weather on the catches of whales. *Rep. Int. Whaling Comm.*, No. 14: 87–91.
Houghton, R. G. and S. Flatman (1980) The exploitation pattern, density-dependent catchability and growth of cod (*Gadus morhua*) in the West-central North Sea *J. Cons. Int. Explor. Mer.*, **39**: 271–287.
ICES (1974) Report of the Working Group on standardization of scientific methods

for comparing the catching performance of different fishing gears; and, procedure for measurement of noise from fishing vessels. *Coop. Res. Rep. ICES*, No. 38: 30 pp.
Kennedy, W. A. (1951) The relationship of fishing effort by gill-nets to the interval between lifts. *J. Fish. Res. Board Can.*, **8** (4): 264–274.
Ketchen, K. S. (1964) Measures of abundance from fisheries for more than one species. *Rapp. P.-V. Reun. CIEM*, **155**: 113–116.
Kimura, D. K. (1980) Standardized measures of relative abundance based on modelling log (c.p.u.e.) and their application to Pacific Ocean perch (*Sebastes alutus*) *J. Cons. int. Explor. Mer*, **39**: 211–218.
Loh Lee Low (1974) Atlas of Japanese fisheries in the north-east Pacific, January 1970–October 1972: Three-dimensional graphs of monthly catch statistics. Seattle, NOAA/NMFS, Northwest Fisheries Center, Processed Report, 95 pp.
Munro, J. L. (1974) The mode of operation of Antillean fish traps and the relationship between ingress, escapement, catch and soak. *J. Cons. int. Explor. Mer*, 35 (3): 337–350.
Murphy, G. I. (1960) Estimating abundance from long-line catches. *J. Fish. Res. Board Can.*, **22** (1): 33–40.
Newman, G. G., R. J. M. Crawford and O. M. Centurier-Harris (1978) The effect of vessel characteristics and fishing aids on the fishing power of South African purse seines in ICSEAF Division 1.6. *Collect. Sci. Pap. ICSEAF*, **5**: 123–144.
Pella, J. J. (1969) A stochastic model for purse-seining in a two-species fishery. *J. Theor. Biol.*, **22**: 209–226.
Pella, J. J. and C. T. Psaropulos (1975) Measures of tuna abundance from Purse-seine operations in the eastern Pacific Ocean, adjusted for fleet-wide evaluation of increased fishing power, 1960–1971. *Bull. I-ATTC*, **16** (4): 283–400.
Pope, J. A. (1975) Measurement of fishing effort. A special meeting held at Charlottenlund Slot, Charlottenlund, 25 and 26 September 1973. *Rapp. P.-V. Reun. CIEM*, **168**: 102 pp.
Robson, D. S. (1966) Estimation of relative fishing power of individual ships. *Res. Bull. ICNAF*, No. 3: 5–14.
Rothschild, B. J. (1972) An exposition on the definition of fishing effort. *Fish. Bull. NOAA/NMFS*, **70** (3): 671–699.
Rothschild, B. J. (1977) Fishing effort. In *Fish Population Dynamics*, J. A. Gulland (ed.). London, Wiley, pp. 96–115.
Sanders, M. J. and A. J. Morgan (1976) Fishing power, fishing effort, density, fishing intensity and fishing mortality. *J. Cons. CIEM*, **37** (1): 36–40.
Sissenwine, M. P. and E. W. Bowman (1978) An analysis of some factors affecting the catchability of fish by bottom trawls. *Res. Bull. ICNAF*, No. 13: 81–87.
Skud, B. E. (1972) A reassessment of effort in the halibut fishery. *Sci. Rep. IPHC*, No. 54: 11 pp.
Skud, B. E. (1978) Factors affecting longline catch and effort. *Sci. Rep. IPHC*, No. 64: 66 pp.
Treschev, A. I. (1975) Fishing units measures. *Rapp. P.-V. Reun. CIEM*, **168**: 34–37.
Ulltang, Ø. (1977) Methods of measuring stock abundance other than by the use of commercial catch and effort data. *FAO Fish. Tech. Pap.*, **176**: 23 pp.

2.4 Surveys

Alverson, D. L. (1971) Manual of methods for fisheries resource survey and appraisal. Part 1. Surveying and charting of fisheries resources. *Fao Fish. Tech. Pap.*, No. 102: 80 pp.
Alverson, D. L. and W. T. Pereyra (1969) Demersal fish exploration in the north-eastern Pacific Ocean; an evaluation of exploratory fishing methods and analytical approaches to stock size and yield forecasts. *J. Fish. Res. Board Can.*, **26** (8): 1985–2001.
Brown, B. E. *et al.* (1976) The effect of fishing on the marine finfish biomass in the

Northwest Atlantic from the Gulf of Maine to Cape Hatteras. *Res. Bull. ICNAF*, No. 12: 49–68.

Burczynski, J. (1979) Introduction to the use of sonar systems for estimating fish biomass. *FAO Fish. Tech. Pap.*, No. 191: 89 pp.

Doi, T. (1974) Further development of whale sighting theory. In *The Whale Problem*, W. E. Schevill (ed.). Cambridge, Mass., Harvard University Press, pp. 359–368.

Edwards, R. L. (1968) Fishery resources of the North Atlantic area. In Fishery resources of the world. *Univ. Wash. Publ. Fish.* (*New Ser.*), No. 4: 52–60.

FAO/UNDP (1975) FAO Regional Fishery Survey and Development Project, Doha (Qatar). *Report of the* Ad Hoc *Working Group on Survey Technique and Strategy*, Rome and Doha, 15–26 September 1975. Rome, FAO, FI:DP/REM/71/278/1: 45 pp.

Forbes, S. T. and O. Nakken (1972) Manual of methods for fisheries resource survey and appraisal. Part 2. The use of acoustic instruments for fish detection and abundance estimation. *FAO Man. Fish. Sci.*, No. 5: 38 pp.

Gulland, J. A. (1975) Manual of methods for fisheries resource survey and appraisal. Part 5. Objectives and basic methods. *FAO Fish. Tech. Pap.*, No. 145: 29 pp.

Mackett, D. J. (1973) Manual of methods for fisheries resources survey and appraisal. Part 3. Standard methods and techniques for demersal fisheries resource surveys. *FAO Fish. Tech. Pap.*, No. 124: 39 pp.

Saetersdal, G. (1973) Assessment of unexploited resources. *J. Fish. Res. Board Can.*, **30** (12), Pt. 2: 2010–2016.

Saville, A. (1977) Survey methods of appraising fishery resources. *FAO Fish. Tech. Pap.*, No. 171: 76 pp.

EXERCISES

Exercise 2.1

In the western part of a fishing area 2000 fish were tagged. Does the fact that 170 tagged fish were recaptured in the western half, and 2 in the eastern half tell much about possible seperation? (i) Is the information about possible stock structure different if the 2 fish in the eastern half were caught more than six months after tagging, and of the 170 in the west (a) 160 were caught within six months and 10 later, or (b) 100 within six months and 70 later? (ii) In subsequent experiments the results are given in Table 2.1.

Table 2.1

		Recaptured in west			Recaptured in east		
Time at liberty (months)		0–6	6–12	12+	0–6	6–12	12+
Tagged in western part	15000	1220	325	74	5	44	32
Tagged in eastern part	3000	14	31	16	61	22	6

Is there any evidence that there are separate stocks in the eastern and western area? [Hint: Calculate the percentage recaptured after 6 and 12 months, and compare the values for a given area from tagging experiments in different areas.]

Exercise 2.2

Table 2.2 gives catch (in tons) and effort statistics (in days fishing) in three successive years, in which data from four size classes of trawlers have been distinguished.

Table 2.2

	1973		1974		1975		1976	
Tonnage class	Catch	Effort	Catch	Effort	Catch	Effort	Catch	Effort
10–39	150	800	50	400	45	300	20	200
40–59	330	600	200	700	250	600	120	500
60–79	290	400	260	600	330	600	160	500
80–99	170	200	160	300	365	500	380	800
Total	940	2000	670	2000	990	2000	680	2000

(a) Calculate the c.p.u.e. of each tonnage class in each year, and the c.p.u.e. of the fleet as a whole. Express the value for each year as a percentage of 1973.
(b) Using the data for 1973, estimate the average fishing power of each class, relative to the 40–59 ton class. Hence, calculate the total effort in each year in units of days fishing by the 40–59 ton class. Calculate the c.p.u.e. in each year (in 40–59 ton units) by dividing this total effort into the total catch.
(c) Does the value for 1973 differ from the c.p.u.e. of the 40–59 ton class? Would you expect it to be different?
(d) What are the best estimates of the changes in fishing mortality and stock abundance, expressing the values in each year as percentages of 1973?

Exercise 2.3

Table 2.3 gives the catch and fishing effort in each season for three different years. Estimate the abundance in 1970 and 1975 as percentages of the abundance in 1965: (a) using the data for total catch and effort data for each year; (b) by calculating the c.p.u.e. in each season each year, and expressing the values in 1970 and 1975 as percentages of the corresponding season in 1965; (c) by calculating the average of the seasonal c.p.u.e. in each year. What is the best estimate of the percentage increase in the real intensity of fishing (fishing mortality) from 1965 to 1970 and 1975?

Table 2.3

	1965		1970		1975	
	Catch	Effort	Catch	Effort	Catch	Effort
Spring	500	5	780	30	480	10
Summer	400	5	210	10	400	10
Autumn	600	15	45	5	210	10
Winter	300	5	75	5	290	10
Total	1800	30	1110	50	1380	40

Exercise 2.4

Table 2.4 is an extreme simplification of United Kingdom North Sea trawl data and represents the effort and catches of plaice and haddock in two years. The area has been divided into 16 subareas in each of which the data of catch and effort have been recorded separately.

(a) For each year, obtain, by addition, total catch of each species and the total effort, and hence calculate the ratio total catch/total effort.
(b) Draw up a chart for each year showing the c.p.u.e. of each species in each rectangle.
(c) Calculate for each species an overall density index, i.e. the mean c.p.u.e., and the effective fishing intensity on each species each year.

Compare the fishing intensities on the two species.

Table 2.4

	Year 1				Year 2			
Effort	5	6	6	3	16	17	13	14
Haddock	50	48	60	24	208	238	195	168
Plaice	0	12	6	0	0	17	13	0
Effort	8	7	9	8	13	12	13	10
Haddock	40	49	54	48	130	132	91	80
Plaice	16	0	27	8	13	12	26	0
Effort	10	13	11	14	9	9	(8)	6
Haddock	40	65	33	56	45	63	(32)	48
Plaice	40	39	22	42	18	18	(8)	18
Effort	14	15	16	15	5	5	6	4
Haddock	28	0	16	15	10	5	6	4
Plaice	84	90	48	45	25	15	12	8

Compare the change in density between the two years as measured by the ratio total catch/total effort, and by mean c.p.u.e.

(d) Supposing that there had been no fishing in the second year in one of the middle rectangles, as indicated by brackets in Table 2.4, how could mean density or effective overall fishing intensity be calculated? Make some reasonable assumption as to the density in the subarea; try the effect of different assumptions. Some assumptions are: that the density is the mean of surrounding subareas; that the change from the previous year is the same as that for other subareas; (as a limiting case) that the density is zero.

Exercise 2.5

Table 2.5 gives the catch and effort statistics for the cod fishery in the International Council for the Exploration of the Sea Region 1 (Barents Sea). The catches are given in tons (the total includes German and Norwegian catches), United Kingdom fishing effort as millions of ton-hours (hours fishing × mean tonnage of the fishing vessels), and USSR fishing effort as thousands of hour's fishing.

Calculate the c.p.u.e. of the United Kingdom and USSR fleets. Calculate the total fishing effort in United Kingdom and in USSR units.

For each fleet express the annual c.p.u.e. as a percentage of the 1946–56 average; are the trends in the two series the same? Might the difference be accounted for by the fact that one series contains a factor (tonnage) which makes some allowance for increased fishing power of the individual trawlers?

Table 2.5

Year	Catch			Effort	
	United Kingdom	USSR	Total	United Kingdom	USSR
1946	53 835	117 100	199 640	17.6	104
1947	127 242	151 970	340 758	38.4	149
1948	164 794	158 650	406 620	63.1	162
1949	226 450	162 340	484 942	80.0	171
1950	136 790	135 410	356 474	93.2	161
1951	129 030	189 580	407 989	98.9	231
1952	130 546	258 830	524 160	102.6	247
1953	59 445	261 400	442 839	53.1	275
1954	72 347	404 650	597 534	51.5	340
1955	91 379	530 280	830 694	60.6	373
1956	67 787	512 170	787 070	54.3	492
1957	38 488	183 000	399 595	44.5	—
1958	46 225	146 570	388 067	55.6	—

Exercise 2.6

German trawlers fish at West Greenland for both cod and redfish. Their catches in tons and fishing efforts during 1958 and 1959 are given in Table 2.6 (data taken from *Statistical Bulletins* of the International Commission for the Northwest Atlantic Fisheries).

Table 2.6

Year	Primary species	Days fished	Catch of cod	Catch of redfish
1958	Cod	1337	26 247	1 754
	Redfish	385	1 277	9 457
	Mixed	199	2 386	1 969
	Total	1921	29 910	13 180
1959	Cod	645	12 336	1 087
	Redfish	690	2 705	15 683
	Mixed	169	2 372	2 062
	Total	1504	17 413	18 832

Estimate, from the c.p.u.e. of cod by the trawlers fishing for cod, and of redfish by trawlers fishing for redfish, the changes (as percentages) in the densities of the stocks of cod and redfish between 1958 and 1959. Compare these with the changes in the c.p.u.e. of cod and redfish by all vessels taken together, and also with the changes in the c.p.u.e. of cod by the redfish vessels, and of redfish by the cod vessels.

Exercise 2.7

(a) Outline briefly the types of gear used in the fisheries with which you are familiar, and the available measures of effort (distinguishing as appropriate fishing power and fishing time).
(b) What changes in the types of vessel or gear or the methods of using them have occurred which might alter the catchability coefficient corresponding to these measures of effort?
(c) What adjustments could be made to the measures of effort to make them more satisfactory?

Exercise 2.8

A survey was carried out by the RV *Dr. Fridtjof Nansen* off the Arabian Sea coast of Oman, using acoustic and trawl gears. The biomass of small pelagic fish was estimated as 650 000 tons, and of demersal fish as 200 000 tons. If the natural mortality of small pelagic fish can be taken as 0.8 and of demersal fish as 0.5, estimate the potential yield of each group, assuming the stocks were not exploited at the time of the survey.

If the demersal biomass was composed of 20 per cent catfish, 20 per cent groupers and other large commercially valuable species, and 60 per cent ponyfish and other small, less valuable species, and the natural mortalities of these groups may be estimated as 0.15, 0.25, and 0.7 respectively, estimate the potential yield of each group and of total demersal. How does the percentage composition of the potential yield differ from that of the original biomass?

If these were actually catches of 50 000 tons of small pelagic fish, and 15 000 of groupers and similar species, how would the estimates be changed?

CHAPTER 3

Production models

3.1 SIMPLE MODELS

Production models form one of the two groups of models used in studying fish population and assessing the state of fish stocks. Unlike the analytic models discussed in later chapters they do not consider the events within a population, and in particular ignore the growth and mortality of the individuals forming the population. Essentially they are concerned with four basic quantities—the population biomass B, the catch Y, the amount of fishing (usually expressed as fishing effort f), and the net natural rate of increase. While this appears to be a very simplistic approach to studying the population it may not be unduly so. At least in principle these models consider all the effects on the net rate of increase (the main dependent variable in most formulations) of changes in population biomass or in amount of fishing, both direct and indirect (e.g. through effects on the stocks of the competitors with, predators on, or food of, the fish stock of interest). In that sense the models are, to follow the jargon often used, 'ecosystem' models. However, this principle holds only if the system has settled into equilibrium corresponding to the particular level of biomass or amount of fishing. In practice, conditions are continually changing, and many of the indirect effects can take some time to work through the system and become fully effective and apparent. As discussed in later sections, considerable attention therefore needs to be paid to consideration and methods of analysis of non-equilibrium situations.

The simplest approach to applying a production model has already been outlined in Chapter 1. Using this approach, we note that what is of immediate practical interest is the relation between the amount of fishing and the catch. Therefore, before attempting more sophisticated analyses, the most direct analysis to attempt is to consider the empirical relation between catch and amount of fishing, i.e. the catch Y, should be plotted against the fishing effort f, and the relation between them examined. That is, the first task is to examine the relation

$$Y = Y(f) \tag{3.1}$$

In practice it is usually more convenient to consider the relation between the catch per unit effort (c.p.u.e.) U (i.e. the stock abundance, if effort has been measured properly) and fishing effort, i.e.

$$U = U(f) \tag{3.2}$$

Clearly the abundance (c.p.u.e.) will decrease with increased fishing, and the simplest assumption of how it will decrease is that the decrease is linear, i.e.

$$U = U(f) = a - bf \tag{3.3}$$

which gives the catch, as a function of effort as

$$Y = af - bf^2 \tag{3.4}$$

Another approach is to note that under common assumptions of how natural populations (whether of fish, or insects, etc.) behave, then in the absence of any exploitation, a population whose biomass B is less than B_∞, the 'carrying capacity' of its environment (i.e. the maximum equilibrium population biomass to which the population will approach in the absence of interference), will tend to increase. The rate of increase will be some function of the current biomass, i.e. we can write

$$\frac{dB}{dt} = f(B) \tag{3.5}$$

and, since $f(B) = 0$ when $B = 0$, and when $B = B_\infty$ the simplest equation that satisfies these conditions is

$$\frac{dB}{dt} = rB(B_\infty - B) \tag{3.6}$$

This equation can also be written in a form which describes the change during a finite period (say a year), as a function of the average biomass $\bar{B}$ during that period, i.e.

$$\Delta B = r\bar{B}(B_\infty - \bar{B})$$

or, if there has been a catch Y during that period, the total change in population will be given by

$$\Delta B = r\bar{B}(B_\infty - \bar{B}) - Y \tag{3.7}$$

and if the catch is adjusted so that there is no change in population size, then

$$Y = r\bar{B}(B_\infty - \bar{B}) \tag{3.8}$$

Equation (3.8) is therefore the equation describing the equilibrium relation between catch and population biomass, under the simple assumptions made (Schaefer, 1954). It can be rewritten in terms of fishing effort, f, by writing the mean biomass $\bar{B}$ as $\bar{B} = 1/q \cdot U$, where U is the c.p.u.e. $= Y/f$. Then equation (3.8) becomes

$$Y = r\frac{1}{q} \cdot U\left(B_\infty - \frac{1}{q}U\right)$$

or

$$q^2 f = r(qB_\infty - U)$$

or rearranging terms

$$U = Y/f = qB_{\infty} - q^2/rf \tag{3.9}$$

and

$$Y = qB_{\infty}f - q^2/rf^2 \tag{3.10}$$

Apart from the value of the constants, equations (3.3) and (3.9) and also (3.4) and (3.10) are identical. That is, the simplest assumptions, either of the general nature of population growth (equation 3.6), or of the empirical relation between c.p.u.e. and effort (equation 3.3) lead to the same interrelations between catch, effort, and population size. That is, the c.p.u.e.—or population size—is linearly related to the effort, and the equilibrium catch is a parabolic function of both effort and population size.

A further characteristic can be derived from equation (3.8) and that is that the greatest value of Y occurs when $B = \frac{1}{2}B_{\infty}$, and is equal to

$$Y_{max} = \tfrac{1}{4}rB_{\infty}^2$$

The point at which the greatest catch (the maximum sustainable yield, MSY), occurs, and the value of this yield, has received considerable attention in the literature concerned with the theory of resource management, especially of marine mammals, but its practical importance is probably not so great (e.g. Gulland, 1968, Larkin, 1977). There are several reasons for this. Relevant to the present discussion is that real fish stocks do not fit the simple models described by the equations above. The curve of catch as a function of population abundance (or effort) usually does not have such a sharply defined maximum as the parabola corresponding to the simplest assumption nor does the maximum always occur tidily at half the unexploited population. For many fish it seems that the maximum catch occurs at a population less than half the unexploited population—e.g. around 0.4 for North Atlantic cod (Garrod, 1969), while for marine mammals it may occur above that level.

3.2 NON-LINEAR MODELS

More realistic expressions for the relation between catch, effort, and population size under equilibrium conditions can be obtained by using non-linear expressions for $U(f)$ in equation (3.2), or for $f^i(B)$ (where $f(B) = Bf^i(B)$, so that $f^i(B)$ is the rate of increase as a proportion of the population) in equation (3.5).

In the former case it has been found that, examining data over a wide range of fishing effort, the c.p.u.e. for many stocks appears to decline exponentially (rather than linearly) as a function of effort. This suggests (Garrod, 1969, Fox, 1970) a relation of the form

$$U = U_{\infty}e^{-bf}$$

or

$$\text{Log}_e U = a - bf \tag{3.11}$$

and with suitable algebraic juggling the relation between population size and rate of change of population (or surplus production), which is equal to the catch in the equilibrium state, is given by

$$Y = \Delta B = \bar{B}k(\log_e B_\infty - \log_e \bar{B})$$

With further algebra it can be shown that the maximum catch C_{max} is given by

$$Y_{max} = U_\infty / be$$

and will occur at a population size B_∞/e (i.e. about $0.37B_\infty$) and at a fishing effort of $1/b$.

Another common generalization of the linear relation is to write equation (3.6), which describes the growth of the population, in the form

$$\frac{dB}{dt} = rB_\infty B - rB^m \tag{3.12}$$

which, with somewhat different constants, is the General Production (GENPROD) model of Pella and Tomlinson (1969). If $m = 2$, equation (3.12) is identical to (3.6), so that the Schaefer model appears as a special case.

The Schaefer and exponential models have only two degrees of freedom—basically the maximum population size, B_∞, and the rate at which the population tends to return to the population if reduced, determined by r or k. However, the GENPROD model has three. The third (m) determines the shape of the yield–effort curve. If $m = 2$ this is a parabola. Values of $m < 2$ correspond to progressively flatter curves, while for $m > 2$ the decrease beyond the level of effort giving maximum catch becomes sharper, tending, for large m, to correspond to sudden and complete collapse. This third degree of freedom thus allows the models to be adjusted to a wider range of actual events, and therefore to be potentially more realistic. At the same time, because of this greater flexibility, the chances of getting a spurious fit to a limited number of data points is increased (see section 3.4). Modifications to the formulation of the Pella—Tomlinson model (Fletcher, 1978b), and to the method of fitting (Rivard and Bledsoe, 1978) have been proposed that reduce the probability of getting spurious fits.

3.3 NON-EQUILIBRIUM SITUATIONS

The equations derived in the previous sections, except for those describing the rate of growth of the population (i.e. 3.5, 3.6 and 3.12), refer to equilibrium, steady-state conditions. The fishery scientist is seldom lucky enough to assess a fishery in which the fishermen have been sufficiently obliging to maintain the fishing effort at the same level over a long enough period for the stock to have settled down to the equilibrium level of abundance corresponding to that effort level. Still less is it likely that the fishery has been maintained at several different equilibrium levels over the years, so that the form of the equilibrium curves can be determined. For practical application, the equations of the previous section have therefore to be modified to take account of non-equilibrium conditions.

The simplest situation is when, in accordance with the simple theory, the growth rate of the population at any moment is directly and uniquely determined by the magnitude of the current population and is not influenced by the earlier history of the population, or by its composition. Then equations (3.6) and (3.12) will be valid, or, considering the change during a year, and taking account of the catch Y during the year, equations (3.7) and the corresponding form of equation (3.12) can be written as

$$B_0 - B_1 = r\bar{B}(B_\infty - \bar{B}) - Y \tag{3.13}$$

or

$$B_0 - B_1 = r\bar{B}\{B_\infty - \bar{B}^{m-1}\} - Y \tag{3.14}$$

where B_0, B_1 = sizes of the population at the beginning and end of the year, and $\bar{B}$ = average biomass during the year. It may be noted in passing that these equations include the implicit assumption that the gross natural increase during the year of a population that is changing is the same as that of a population held constant at the average population size $\bar{B}$, but this is not a very demanding assumption.

In real life only populations of the most primitive animals can be expected to react instantaneously, or even approximately so. For fish there must be some delay in some effects. It may, for example, be several years before the young fish spawned by an unusually large adult stock have grown large enough to produce the corresponding increase in the stock. Other important effects such as the effect of a large population on the supply of food and hence on the growth rate, may have involved different delay periods, but for simplicity following for example Walter (1973), the rate of increase can be assumed to be determined by the current population, and the population at some time T previously. Then equation (3.5) can be rewritten as

$$\frac{dB_t}{dt} = f_0(B_t) + f_T(B_{t-T}) \tag{3.15}$$

While this approach offers some theoretical advantages, in practice the uncertainties about the form of $f_0(B)$ and $f_T(B)$ (which need not be the same, and could correspond to different elements of the instantaneous growth rates of equations (3.6), (3.12), or similar), and the magnitude of the delay T, have not yet allowed this approach to be widely used.

Alternatively, non-equilibrium conditions can be looked at from the point of view of the relation between stock abundance and the amount of fishing. The current abundance will be hardly influenced at all by the level of fishing sufficiently far into the past, but will be affected by more recent fishing. The span of time over which the level of fishing in a particular year affects the stock abundance in later years depends in part on the degree to which the abundance of young recruits entering the fishery is affected by the abundance of the present stock (see Chapter 6). If the abundance of adults, over the range observed in practice, significantly affects the recruitment, then the fishing in any one year can have effects for a period equal to at least one full generation, and probably more, into the future. This can make analysis difficult, and if there is a significant stock/recruit relation, and the fish are long-lived, or the span of years for which these data are small, then it may not be practical to use these types of production model, or at best they should be used with caution. For example see Holden (1977, Figure 30), concerning North Sea dogfish.

Fortunately, these cases are rare. More commonly the number of recruits is effectively independent of the adult stock over most of the observed range of stock size. Then the abundance of a particular age-group in the fishery is affected only by the amount of fishing during the time that cohort has been in the fishery. The youngest, newly recruited, fish will have been affected only by the fishing effort in the current year. Older fish will be affected by a longer period of fishing. For example those that were recruited four years previously will have been affected by the effort over those four years. The total stock abundance can therefore be related to some average effort over a period that is no longer than the greatest life span of an individual fish in the fishery. Suggestions for appropriate weighting factors are made by Fox (1975) but, as he notes, the correct factors will vary as changes in fishing affect the average life span—at low rates of fishing, many old fish will be present, and their abundance will be affected by fishing some time in the past. Using different weighting factors for each year can be laborious, and might suggest quite false degrees of precision. Better is to use a simple averaging procedure. A common one is to take the average effort over a period T approximately the *average* number of years that an individual fish spends in the fishery. This will be perhaps a half to a third the maximum life span, e.g. if a fish recruit at 3 years old, and 10-year-old fish are the oldest found in the fishery, the c.p.u.e. can be related to the average fishing effort over 3 years (the current year and the two previous). That is, equation (3.2) can be written, for the c.p.u.e. in year t,

$$U_t = U(\bar{f}_t)$$

where (3.16)

$$\bar{f}_t = \frac{1}{T}\sum_{i=0}^{T-1} f_{t-i}$$

3.4 APPLICATIONS TO DATA

3.4.1 General considerations

If the quantities involved (effort, population, biomass, etc.) could be estimated without error, and the events in the population followed exactly one or other of the models outlined above without other sources of variation (fluctuations in year-class strength, etc.) having any effect, then applying the models to observed data would not present problems other than selecting the appropriate model. In practice the result of the many sources of variation on the observed data is that the data can be fitted to the models in various ways, with a consequent range of estimates of values of the parameters in the operations, and of the important population characteristics, e.g. population size or fishing effort giving the greatest yield, etc. It is therefor desirable to use a method of fitting one or other model to the observed data that makes the most effective use of the observations, and that is both likely to result in useful advice and unlikely to result in misleading advice.

One approach to fitting has been to manipulate the basic data (usually the annual catch and effort statistics) in such a way that the procedure becomes a matter of fitting a simple curve to pairs of values derived from these data, for example the c.p.u.e. in a given year, and the average of the effort in that year and in the previous year. At its simplest and least sophisticated, the final fitting can be done by drawing some appropriately chosen curve through the points by eye.

At the opposite extreme a computer program can be written that calculates the expected value of the catch (or of the c.p.u.e., or some other feature of the fishery), as a function of the observed characteristics of the fishery (especially the current and previous levels of fishing effort) and of the stock and its environment (e.g. temperature at the time the dominant age-group in the fishery was spawned), on the basis of the preferred population model. The computer program then finds which set of population parameters (e.g. the values of r, m and B_∞ in equation 3.14) gives the best fit between observed and expected values, usually as a minimum sum of squares of differences—for example min $(Y_o - Y_E)^2$.

As usual the wise man will find a middle course between these extremes that uses the advantages and avoids the disadvantages of each approach. A good statistical procedure will provide an estimate that is 'best' in some definable sense, for example which results in the least mean-square deviation between observed catches and those predicted on the basis of the model used, and the parameters that have been estimated. It should be remembered that just as there are many ways in which a 'best' management procedure can be defined, so there are many ways in which the 'best' fit can be defined. For example, minimizing the sums of squares between observed and expected catches per unit effort might give a quite different 'best' fit from that obtained by minimizing the deviation in catches.

A big advantage of formal statistical or other objective procedures is that their application is indeed objective. Therefore, given the observed set of data, the models, and the fitting procedure, the resultant estimates of parameters will be the same (arithmetical errors permitting) whoever does the fitting. This can be most useful when negotiations between varied interests are going on, when arguments on fitting curves can hold up discussions on other matters. Another potential advantage of objective procedures is that, with suitable computer programs they can enable complex relations including several parameters and non-linear functions to be fitted. Some of these advantages may be more apparent than real. When the number of data points are few, and the fitting procedure complicated, errors in estimating the basic data (e.g. total effort), or natural variations in the system not described by the model can result in estimates of parameters that give the 'best' fit differing widely from the true value, and subject to considerable changes when extra data points, e.g. from another fishing season, become available. Certain parameters are especially subject to this difficulty. For example estimates of q (catchability coefficient) obtained from a least-squares fit of the GENPROD model, which also simultaneously estimates m and B_∞ (equation 3.12), can imply, from the actual values of effort, quite unrealistic values of fishing mortality. In practice it is often preferable, when fitting the GENPROD model to fix q (or set boundaries to its possible values). Also, there may be a feeling among the unwary users that because a certain set of values of population parameters has been derived from a sophisticated fitting procedure these values must therefore be correct.

The practical solution is to use whatever objective fitting procedure or computer procedure appears most appropriate, but always check on the reliability of the result, and the alternative fits. This is best done by arranging that the fit can be demonstrated graphically, e.g. as a plot of population size as a function of amount of fishing effort with the points corresponding to each year's data shown. The latter is important because as well as suggesting what year (or years) do not fit the model (and therefore indicating when the data may be suspect, or equally likely, whether some extra influence, not considered by the model is having a significant effect on the stock), the plot will also show which years are having a major influence in the particular parameter estimates that have been obtained. It often happens that in one year exceptional values of say, c.p.u.e. and effort occur, and this pair of values can have a great influence on the values of parameters of the model to which the data are fitted. Data for such a year should be treated with as much suspicion as those for a year that fails to fit.

Care should be taken in using standard statistical techniques because the data may not wholly satisfy the basic assumptions. For example, in fitting a regression of c.p.u.e. on effort, the independent variable (effort) is often not known accurately and can be subject to errors in various steps in estimation (e.g. in making allowance for changes in the fishing power of the vessels). Then the usual least-squares regression will not reflect the functional relation between the actual fishing effort and c.p.u.e., but will provide a prediction of the expected c.p.u.e. corresponding to a given estimated fishing effort (see Ricker, 1972; Jolicoeur, 1975; Ricker, 1975b). The latter will tend to result in a flatter relation between the two variables (i.e. underestimate the reduction in c.p.u.e. caused by a given increase in effort).

Another possible source of error, occuring when relating c.p.u.e. to effort, which tends to cause an opposite bias, arises from the fact that c.p.u.e. and effort are usually not independent, but one is derived from the other by dividing into the total catch. If catch were constant, the procedure would be equivalent to relating $1/x$ to x, which gives a negative correlation (strictly a hyperbola). The possibility of producing a wholly artificial correlation in this way, unconnected with any real effect of fishing on the stocks has often been noted.

However, quantitative measures of the extent of the effect possible have seldom been calculated, nor have there been many examples where these potential errors have actually resulted in biased interpretations of specific situations. In practice it may be reasonable to ignore the possible bias provided that the likely errors and variations in estimating the value of the fishing effort in any particular year is small compared with the range of values of the actual effort during the period of analysis. The question of the bias introduced has been examined in detail in a number of recent papers (e.g. Uhler, 1980; Mohn, 1980).

An alternative method of avoiding the introduction of a false correlation caused by the same variable appearing in both sides of the relation, is to relate the c.p.u.e. in year x to the effort in year $x-1$ (Walter, 1975). This may also be realistic if there is a lag in the effect of fishing in the stock. Walter also points out that assuming the standard logistic growth model of equation (3.6) holds, the regression line fitted to the pair of points (U_x, f_{x-1}) will not give the exact equilibrium relation corresponding to that model, and suggests further adjustments. Since these adjustments are valid only if the logistic model is satisfied, and fish populations seldom grow in such a simple fashion, these additional refinements are probably not worth pursing.

Wrong conclusions about the effect of fishing on the stock, and the possibilities of increasing catches by fishing harder can also arise if the amount of fishing is affected by the abundance or availability of the stock. This can easily happen if the fishery operates on several stocks (different species, or the same species in different areas), and can switch from stock to stock according to which is the more abundant. In such a situation, if fishing is having no effect on the stock, one could expect a positive correlation between observed effort and stock abundance, i.e. higher c.p.u.e. at higher levels of effort. If fishing is having an effect, the net result may be a negative correlation, but in any case the effect, the fishing, i.e. the reduction of c.p.u.e. by fishing, will be underestimated.

To some extent this situation can also be treated by relating the c.p.u.e. in year x to the effort in year $x-1$. This will be most effective if the natural fluctuations on the stocks are in availability (i.e. the catchability coefficient q) and are of sufficiently short duration that a high q in year $x-1$ (causing a high effort) will not be correlated with a high q in year x (i.e. the c.p.u.e. will be as likely to be low as high). If there are real changes in abundance (e.g. good or bad year classes) this technique is less effective. For a very short-lived fish, the effort in year $x-1$ will have little effect on the c.p.u.e. in the next year. For longer-lived fish, good or bad year-classes will affect the abundance (and hence the effort) over several years.

It seems that the possibility of being misled by this type of relation cannot be removed merely by arithmetical juggling, and whenever the amount of fishing in one year may increase because catches are particularly good in that year, any catch/effort analysis should be interpreted with care. Thus, before making (or at least before using) the analysis, these questions should be asked:

(i) does the fishing effort vary much from year to year, i.e. as opposed to showing a constant trend over a long period?

(ii) are any of these changes due to effort increasing when catches are good, especially through diversion from other stocks?

3.5 FITTING PROCEDURES

The practical procedures that should be followed in fitting a production model to actual data can now be reviewed. They can be arranged according to the basic relation used in fitting—c.p.u.e. as a function of fishing effort, or natural rate of change as a function of popultion. Other possible relations that could be used are less suitable. For example, catch could be examined as a function of effort, but this relation is so dominated by the direct proportional relation between current effort and current catches that it is difficult to determine the underlying effect of current and earlier fishing on the stock.

The first step is to assemble the data of catch, effort, and c.p.u.e., making sure that all catches from all fisheries on the stock are included, even if the data on effort or c.p.u.e. only come from one group of vessels. These should then be examined to see whether the bias mentioned in section 3.4 is likely to occur. That is, whether a significant cause of variation in the dependent variable might be random errors in the estimation of the independent variable, for example whether high values of estimated c.p.u.e. are as likely to occur because the estimated value of effort is low, as because the population density is really high. If this is the case the c.p.u.e. in year x should be plotted as a function of the effort in year $x - 1$. Otherwise the c.p.u.e. can be plotted directly as a function of the effort in the same year. These plots should give some immediate insight into the effects of fishing on the stock.

The next stage is to consider what lags there might be in the system, i.e. the degree to which past levels of fishing effort affect the current stock abundance. The most important question is whether the level of recruitment of young fish to the fishery is significantly influenced by the abundance of adults (see Chapter 6). If it is, then the amount of fishing in one year could affect the abundance on one or two generations (and hence perhaps 10 years or more) later. Analysis by the methods of this chapter would then be difficult. In practice, if enough is known about the population and its structure and dynamics to be confident that changes in adult stock affect recruitment, then there would be enough information to use the more complex analytical methods of analysis described in the following chapters. The practical question is whether there might be reason to suspect that the fishery has, by affecting the adult stock, caused the recruitment to fall to any significant extent. If so, analysis should proceed carefully. So also should any development in the fishing, since the effects of heavy fishing could take some time to become apparent. A direct clue that recruitment might be being affected comes from the size of fish caught. If recruitment stays constant under increasingly heavy fishing, the average size of fish will fall. Conversely, if the average size of individual fish remains the same while the abundance, or c.p.u.e., falls, there is some evidence that recruitment is being affected.

Less direct evidence can be obtained from the biology of the fish. If fecundity is low, then the total mortality of eggs, larvae, and young fish before recruitment must, on the average, also be low. It is therefore less easy for this mortality to decrease sufficiently when adult stock is low to make up for the reduction in the initial size of the brood. Significant density-dependent recruitment is therefore more likely in fish with low fecundity, notably elasmobranchs.

If recruitment is not affected, then fishing in one year can affect the stock only so long as some fish exposed to fishing in that year remain in the exploited stock. This potential life span in the fishery, of say L years, provides an upper limit to the length of period that needs to be considered. Most fish will remain in the fishery for a much shorter period; or, viewed another way, most fish in the fishery at a given moment have been exposed to a much shorter period of fishing. As a reasonable approximation the mean life span in the fishery may be taken as half to a third of the potential life span and the mean effort calculated over that period. A more exact formula for computing the average effort is given by Fox (1975), as

$$\bar{f} = \{(lf_x + (l-1)f_{x-1} + \dots f_{x-1} + 1\}/\{l + \dots + 1\}$$

The c.p.u.e. in each year should then be plotted graphically against the mean effort corresponding to that year. This plot will first show how good the relation is; that is, to what extent the fishing effort is determining the c.p.u.e. Second, it will suggest what form of equation describing the relation is the most reasonable (a straight line, exponential, etc.). Appropriate fits can be made, e.g. normal linear regression of c.p.u.e. on f; of log (c.p.u.e.) on f; or the PRODFIT computer program (Fox, 1975) if the GENPROD model is believed to give the best fit.

The statistical procedures used in fitting should of course include, so far as possible, tests of significance and calculation of confidence limits on the estimates, but their interpretation needs care, especially when the data are scattered. The statistical procedure is to put forward a null hypothesis, and test whether the data enable this hypothesis to be rejected, i.e. whether, if the hypothesis were true, something unlikely (e.g. something that would occur only 1 per cent of the time) has occurred. The normal null hypothesis for a statistician when considering a relation such as the regression of c.p.u.e. on effort would be to suppose that the two variables are not related. In present-day fisheries where most stocks are at least moderately heavily fished this is an unreasonable hypothesis. Indeed, it is an unwise hypothesis, and in spirit is directly opposed to what should be the attitude of those concerned with managing or administering fisheries. This should be to presume in the absence of other evidence that fishing is having some effect on the stock, and to permit or encourage no more than a moderate expansion of fishing unless it is clear that the stock can support a greater expansion. If a null hypothesis is required, it should be one that supposes that there is some relation between c.p.u.e. and effort. In practice, it is easier to avoid any test of significance at all, and express the statistical variability as confidence limits on the regression lines of c.p.u.e.

on effort. These limits can then be converted to provide ranges in the other relations of interest (e.g. between total catch and effort).

Instead of fitting c.p.u.e. as some function of effort, one can fit catches as a function of effort or population size. For example the GENPROD computer program (Pella and Tomlinson, 1969) predicts catches in each year as a function of the observed effort and the past history of the fisery, and minimizes the difference between observed and predicted catches. This procedure does not seem to differ fundamentally from fitting c.p.u.e., but since the catch in any particular year is very largely determined by the effort, fitting catch may give less insight into what is actually happening to the stock. Also the statistical fit, as given by the correlation coefficient between observed and predicted catch, may give the impression of a better agreement between the model (or the estimates of the parameters in the model) and reality.

Another approach is to fit the changes in population. While this is the original method used by Schaefer (1954), it has not proved as useful as other methods because in most stocks there are large changes in stock size, due, for example, to variations in year-class strength, that have little if anything to do with population size or fishing. One exception is provided by whales (see exercise 3.4), in which the natural changes (recruitment and mortality) do not vary much from year to year and are very closely related to population size (partly the current population, and partly the population a year earlier, where a is the average age at which young whales recruit to the population).

The fitting procedure can be deduced from equation (3.7), rewritten in the form

$$r\bar{B}(B_\infty - \bar{B}) = \Delta B + Y \tag{3.17}$$

or more generally, in terms of the values in year i,

$$f(\bar{B}_i) = \Delta B_i + Y_i$$

where ΔB_i = change in population $= B_{i+1} - B_i$, and B_i = population at the beginning of the ith year.

This would provide a practical method of fitting (calculating $f(\bar{B})$, as a function of catch and population sizes) if point estimates of population sizes were available. Usually only c.p.u.e. is known, and only as an average over a year. To a first approximation, however, we can write

$$B_i = \tfrac{1}{2}\{\bar{B}_i + \bar{B}_{i-1}\}$$

and since

$$\bar{B}_i = \frac{1}{q}\bar{U}_i$$

$$\Delta B_i = \frac{1}{2}\left\{\frac{1}{q}\bar{U}_{i+1} + \frac{1}{q}\bar{U}_i\right\} - \frac{1}{2}\left\{\frac{1}{q}\bar{U}_i + \frac{1}{q}\bar{U}_{i-1}\right\}$$

i.e.

$$B_i = \frac{1}{2}\frac{1}{q}\{\bar{U}_{i+1} - \bar{U}_{i-1}\}$$

My colleagues in FAO, John Caddy and Jorge Csirke, have recently pointed out that the basic theory of production models can be applied even when effort data are missing, provided there is some measure of total mortality (e.g. from age or size data). Equation (3.4) can be rewritten in the form

$$Y = a'F - b'F^2$$

$$= a'(Z - M) - b'(Z - M)^2$$

or

$$Y = -b'M^2 - a'M + (a' - 2b'M)Z - b'Z^2 \qquad (3.18)$$

which is a parabolic function of Z, with an intercept on the axis at $Z = M$. The relation may then be determined by plotting Y against Z, and drawing a parabola by eye.

REFERENCES AND READING LIST

Fletcher, R. I. (1978a) Time dependent solutions and efficient parameters for stock production models. *Fish. Bull. NOAA/NMFS*, **76** (2): 377–388.

Fletcher, R. I. (1978b) On the restructuring of the Pella–Tomlinson system. *Fish. Bull. NOAA/NMFS*, **76** (3): 515–522.

Fox, W. W. (1970) An exponential surplus-yield model for optimizing exploited fish populations. *Trans. Am. Fish. Soc.*, **99** (1): 80–88.

Fox, W. W. (1971) Random variability and parameter estimation for generalized production model. *Fish. Bull. NOAA/NMFS*, **71** (4): 1019–1028.

Fox, W. W. (1975) Fitting the generalized stock production model by least-squares and equilibrium approximation. *Fish. Bull. NOAA/NMFS*, **73** (1): 23–36.

Garrod, D. J. (1969) Empirical assessments of catch/effort relationships in North Atlantic cod stocks. *Res. Bull. ICNAF*, No. 6: 26–34.

Graham, M. (1935) Modern theory of exploiting a fishery and applications to North Sea trawling. *J. Cons. CIEM*, **10**: 264–274.

Gulland, J. A. (1961) Fishing and the stocks of fish at Iceland. *Fish. Invest. Minist. Agric. Fish. Food U.K.* (*Series* 2), **23** (4): 52 pp.

Gulland, J. A. (1968) The concept of the maximum sustainable yield and fishery management *FAO Fish. Tech. Pap.*, No. 170: 13 pp.

Holden, M. J. (1977) Elasmobranchs. In *Fish Population Dynamics*, J. A. Gulland (ed.). London, Wiley, pp. 187–215.

Jolicoeur, P. (1975) Linear regressions in fishery research: some comments. *J. Fish. Res. Board Can.*, **32** (8): 1491–1494.

Joseph, J. and T. P. Calkins (1969) Population dynamics of the skipjack tuna (*Katsuwonus pelamis*) of the eastern Pacific Ocean. *Bull. I-ATTC*, **13** (1): 273 pp.

Larkin, P. A. (1977) An epitaph for the concept of maximum sustainable yield. *Trans. Am. Fish. Soc.*, **106** (1): 1–11.

Marchesseault, G. D., S. B. Saila, W. J. Palm (1976) Delayed recruitment models and their application to the American lobster (*Homarus americanus*) fishery. *J. Fish. Res. Board Can.*, **33** (8): 1779–1787.

Mohn, R. K. (1980) Bias and error propagation in logistic production models. *Can J. Fish. Aquat Sci.*, **37** (8): 1276–1283.

Pella, J. J. and P. K. Tomlinson (1969) A generalized stock production model. *Bull. I-ATTC*, **13** (3): 421–496.

Ricker, W. E. (1972) Linear regressions in fishery research. *J. Fish. Res. Board Can.*, **30** (3): 409–434.

Ricker, W. E. (1975a) Computation and interpretation of biological statistics of fish populations. *Bull. Fish. Res. Board Can.*, No. 191: 382 pp.
Ricker, W. E. (1975b) A note concerning Professor Jolicoeur's comments. *J. Fish. Res. Board Can.*, **32** (8): 1494–1498.
Rivard, D. and L. J. Bledsoe (1978) Parameter estimation for the Pella–Tomlinson stock production model under non-equilibrium conditions. *Fish. Bull. NOAA/NMFS*, **76** (3): 523–534.
Schaefer, M. B. (1954) Some aspects of the dynamics of populations important to the management of marine fisheries. *Bull. I-ATTC*, No. 1: 25–56.
Schaefer, M. B. (1957) A study of the dynamics of the fishery for yellowfin tuna in the eastern tropical Pacific Ocean. *Bull. I-ATTC*, No. 2: 245–285.
South China Sea Fisheries Development and Coordinating Programme (1976) *Report of the Workshop on the Fishery Resources of the Malacca Strait.* Pt 1, 29 March–2 April 1976, Jakarta. Manila, SCS/GEN/76/2: 85 pp.
Uhler, R. S. (1980) Least squares regression estimates of the Schaefer production model. *Can J. Fish Aquat. Sci.*, **37** (8): 1284–1294.
Walter, G. C. (1973) Delay–differential equation models for fisheries biomass. *J. Fish. Res. Board Can.*, **30** (7): 939–945.
Walter, G. C. (1975) Graphical methods for estimating parameters in simple models of fisheries. *J. Fish. Res. Board Can.*, **32** (11): 2163–2168.

EXERCISES

Exercise 3.1

Table 3.1 gives the basic statistical information on the demersal fishery in the Malacca Straits by Indonesia, Malaysia, and Thailand. Available data included the total catch (in thousand tons) by each country, the number of boats in Malaysia, the number of fishing days (in thousands) by Indonesian vessels, and the c.p.u.e. (kg/hour) by Thai survey trawlers using a standard gear.

Table 3.1

	1965	1966	1967	1968	1969	1970	1971	1972	1973	1975
Catch (000 tons)										
Indonesia	—	—	—	—	50	53	44	66	89	82
Malaysia	54	60	84	92	76	76	89	97	127	144
Thailand	16	17	63	91	216	183	186	187	216	209
Effort										
Indonesia	—	—	—	—	307	438	535	581	502	528
Malaysia	1940	1670	1530	1900	2360	3630	4250	5100	4200	4860
C.p.u.e.										
Thailand	—	404	294	215	191	97	116	—	—	—

(i) Calculate the c.p.u.e. in the Indonesian and Malaysian fisheries, and the effort in the Thai fishery (in units of thousands of survey-trawler hours).
(ii) By expressing the c.p.u.e. in each fishery in each year as a percentage of a standard year (e.g. 1969) or otherwise, compare the trends in apparent abundance in each fishery. Are they similar? Which series is more likely to be reliable and free from changes in

fishing power or catchability? Obtain a pooled estimate of a standard index of abundance in each year.

(iii) For each country separately, plot c.p.u.e. against effort and hence estimate the relations between catch and effort. What is the greatest sustained catches that can be taken in each fishery? At what levels of effort can these be taken? How do these levels compare with current levels of effort? What advice should be given to the governments concerned?

Exercise 3.2

Table 3.2. gives the catch per unit effort (catch, in tons, per 1 000 ton-hours' fishing by United Kingdom trawlers) and the total catch, in thousands of tons, of cod and haddock, at Iceland.

Table 3.2

Year	Total catch		Catch/effort		Year	Total catch		Catch/effort	
	Cod	Haddock	Cod	Haddock		Cod	Haddock	Cod	Haddock
1906	105	35	0.97	0.55	1936	280	25	1.14	0.14
1907	115	30	1.03	0.49	1937	295	25	1.39	0.11
1908	125	31	1.20	0.54	1938	305	24	1.11	0.16
1909	115	28	1.05	0.57	1939	198	13		
1910	135	30	1.12	0.56	1940	148	20		
1911	155	31	1.26	0.58	1941	158	15		
1912	160	28	1.20	0.47	1942	176	17		
1913	180	26	1.27	0.38	1943	189	13		
					1944	220	12		
1919	130	35	1.41	1.01	1945	216	14		
1920	190	53	1.39	0.70	1946	237	28	1.84	0.63
1921	175	37	1.16	0.58	1947	246	33	1.46	0.42
1922	220	40	1.27	0.61	1948	293	52	1.27	0.32
1923	220	38	1.09	0.44	1949	315	66	1.17	0.37
1924	300	38	1.09	0.30	1950	321	60	0.98	0.24
1925	320	35	1.28	0.31	1951	327	54	0.94	0.19
1926	295	37	1.09	0.32	1952	392	45	0.91	0.18
1927	340	54	0.99	0.38	1953	515	53	1.11	0.18
1928	360	57	0.88	0.34	1954	546	62	1.01	0.17
1929	390	54	0.83	0.29	1955	537	64	1.04	0.21
1930	490	49	1.10	0.28	1956	482	62	1.02	0.19
1931	475	40	1.09	0.21	1957	453	76	0.81	0.16
1932	480	32	1.34	0.19	1958	511	70	0.80	0.14
1933	510	26	1.28	0.14	1959	454	64	0.68	0.18
1934	480	26	1.14	0.14	1960			0.64	0.18
1935	400	27	1.16	0.14					

(i) Calculate the total fishing effort, in United Kingdom units, on cod and on haddock each year.

(ii) If the average duration of life in the fishery is two years for haddock and three for cod, relate the stock abundance to the past fishing effort. Is there any suggestion that either of these relations has changed? In particular, examine the haddock curve for the years

Table 3.3

	1951	1952	1953	1954	1955	1956	1957	1958	1959	1960	1961	1962	1963	1964	1965
Index of abundance	0.28	0.21	0.34	0.40	0.29	0.49	0.33	0.34	0.30	0.29	0.33	0.36	0.38	0.34	0.37
Effort (000 tons standard days)	—	25.6	25.4	21.1	16.1	20.8	19.4	19.0	13.7	11.1	15.0	15.9	18.4	12.7	15.8

since 1950, when a larger mesh was introduced and trawling was banned in certain areas, including some nursery areas.

(iii) What change in effort, from the level occuring in 1960 would be required to achieve the greatest yield (a) of cod, (b) of haddock, under past-1950 condition, (c) of haddock under pre-1960 conditions?

(*Note*: This may be done graphically, distinguishing points for different periods.)

Alternatively, the PRODFIT or GENPROD computer programs can be fitted separately to each of the periods 1906–14, 1919–38, 1939–49, and 1950–60.

Exercise 3.3

Table 3.3 gives statistics of the index of abundance and fishing effort for skipjack in the eastern tropical Pacific (data from Joseph and Calkins, 1969, Table 7).

Determine the correlation between abundance and effort in the same year. (Plot c.p.u.e. against effort, and show confidence limits on the regression.) What would you conclude about the effect of fishing on the skipjack stock if skipjack were the only objective of the fishery? In fact the main target species is yellowfin tuna, and skipjack are only specifically sought if yellowfin are scarce, or skipjack abundant. What difference does this make to the concluskons? (*Note*: Skipjack are short-lived, and spend no more than a year in the fishery.)

Exercise 3.4

Table 3.4 gives data on catches and c.p.u.e. of blue whales in the Antarctic. Taking alternative values of the population size corresponding to a c.p.u.e. of 1.00 as 20 000 and 30 000 whales, calculate the change in population size each year, and hence the sustainable yield each year. Plot this against population size in that year, and that five years earlier (assuming blue whales recruit to the exploited stock at five years of age).

How may the catch data for the years 1939–46 be used?

What is the maximum sustainable yield, and at what population size is it taken?

Table 3.4

Year	Total catch	Catch per unit effort
1925/26	4 697	3.30
1926/27	6 545	3.76
1927/28	8 334	4.39
1928/29	12 734	4.49
1929/30	17 898	3.13
1930/31	29 410	3.71
1931/32	6 488	4.83
1932/33	18 890	4.61
1933/34	17 349	4.76
1934/35	16 500	3.46
1935/36	17 731	3.96
1936/37	14 304	2.83
1937/38	14 923	2.00
1938/39	14 081	1.77
1939/40	11 480	

Table 3.4 (*Cont.*)

Year	Total catch	Catch per unit effort
1940/41	4943	
1941/42	59	
1942/43	125	
1943/44	339	
1944/45	1042	
1945/46	3606	1.36
1946/47	9192	1.85
1947/48	6908	1.12
1948/49	7625	0.98
1949/50	6182	0.80
1950/51	7048	0.81
1951/52	5130	0.63
1952/53	3870	0.44
1953/54	2697	0.42
1954/55	2176	0.31
1955/56	1614	0.37
1956/57	1512	0.27
1957/58	1690	0.28
1958/59	1187	0.18
1959/60	1228	0.14
1960/61	587[a]	0.105
1961/62	639[a]	0.002

[a] Omitting pygmy blue whales.

CHAPTER 4

Parameter estimation

4.1 INTRODUCTION

This chapter is concerned with estimating the individual elements which describe the different aspects of the population. Chapter 5 describes how they can then be combined to provide an analysis of the population and a description of its reaction to exploitation that gives more insight into what is actually happening than the production models described in Chapter 3. The three elements examined here all concern the history of a group of fish (a cohort, or year-class) as it passes through the fishery from the time the individual fish first become big enough to be caught until the last survivor dies or is caught. They describe the pattern of the entry of the young fish into the fishery (recruitment and selection), the decrease in numbers once they have entered the fishery, and the growth of the individual fish.

Emphasis is given to obtaining quantitative mathematical descriptions of these aspects of the population which can be incorporated into the later equations describing, for example, the yield in weight from a given pattern of fishing. Less attention is given to the biological background of the different processes. The present manual is not concerned with a detailed examination of what are the causes of natural mortality (what eats the fish, or what diseases it suffer from) or how much food it needs to grow at a given rate; the principal concern is with estimating the percentage that dies each year from natural causes, and calculating the actual growth rate. However, the more basic biological studies are of potential interest in stock assessment. A knowledge of metabolic patterns, feeding habits, etc. can help in determining what the growth rate is likely to be; they are likely to be particularly valuable when there are questions of changes in the natural parameters, for example a decrease in natural mortality due to increased exploitation of a predator (see Chapter 6).

4.2 GROWTH

4.2.1 General considerations

Fishermen are usually interested in the weight of fish they catch, not the numbers; indeed the value of a catch of a given weight often increases when the average size of fish in the catch is greater, i.e. when there are fewer individuals. Many other characteristics of the population, e.g. the total number of eggs produced during the spawning season, are more closely related to total biomass than to numbers. Assessment of the events during the life of a cohort of fish is largely a matter of the balance between decreasing numbers of individuals, and increasing individual weight. The important aspect of growth is therefore the growth in weight while in the fishery, and the growth pattern earlier in life, before fish are available to the fishery, is less important. The critical input data relate to the rate of growth; the success of a mathematical growth curve should be judged by the fit of predicted growth rate while in the fishery rather than to the fit of weight (or length) at age over the whole life span. At times the concentration of interest is over an even shorter part of the whole life span. For example, when evaluating the effects of a change in the mesh size used in a trawl fishery the critical factor is the growth during the short period between the time when the fish would have been retained by the original small mesh and when they are big enough to be retained by the larger mesh.

4.2.2 Input data

The analytic approach to fish population dynamics was originated in respect of well-behaved species of temperate waters (plaice, salmon, etc.). These species were obliging enough to lay down clearly recognizable marks on their scales corresponding to annual events such as the check in growth during the winter. By counting these the age of any individual fish can be determined, a process that is made easier by the fact that most of the species concerned have a well-defined and usually fairly short spawning season, so that the ages of all the fish in the population differ from each other by an integral number of years. The process of recording age data is usually further simplified by assigning all fish an arbitrary birth date, normally at the end of the calendar year or near the beginning of the spawning period. Thus, for example, a plaice hatched in February 1976 would be 0-group until 31 December 1976, 1-group during 1977 and so on. From observations on age and size of individual fish it is then easy to prepare a table giving the average size at each age (see Exercise 4.3). Tables of this type are the traditional raw material used in fitting growth curves, and remain the commonest input.

It may be noted that under favourable conditions additional data can be extracted from the rings on scales. By noting the width of the scale at each annual ring, and relating scale width to fish length, the size of the fish at the time of each annual check in growth can be determined. In principle this allows a growth curve to be estimated for each fish, rather than an average curve for the population as a whole. It also allows the growth increment in any particular calendar year to be calculated. These can be valuable weapons, allowing the study of variation in growth, including the effect of differences in environmental conditions between years.

However, not all fish are obliging enough to carry clear birth certificates with them. Improved techniques of handling scales and otoliths can make annual rings easier to read (Holden and Rait, 1974) and in tropical areas, where seasonal cycles are less pronounced, but the fish are shorter-lived, counting of daily rings on otoliths appears to be a promising technique (Brothers *et al.*, 1976). Nevertheless, it has to be expected that for some time to come direct age determination will be difficult to carry out with any reliability for many tropical and subtropical species. Other methods have to be used for these species to obtain input data for fitting growth curves.

The most direct of these methods is tagging. In principle there is nothing simpler or more direct than catching a fish, measuring it, tagging and releasing it, and then waiting until it has been recaptured, and recording its growth during the time at liberty. The practical difficulties are that the operation of catching, handling, measuring, and tagging the fish may give it a shock, which it may take some time to recover from, and during this time growth will be below average. At the extreme some fish, e.g. tuna, are so tender that it is unwise to keep them long enough to measure; in these cases a fair estimate of the size when tagged can be obtained from the size of similar sized fish that were not tagged. On the other hand, tougher fish can be injected with a substance such as tetracycline (e.g. Holden and Vince, 1973) which will be laid down as a visible ring on the scale or otolith, and thus be a welcome aid in later interpretations when the fish is recaptured.

Another method of value when the fish has distinct breeding seasons is the Petersen method. When conditions are favourable, i.e. when the fish grow quickly and uniformly, without much difference between individuals, this is very simple. If the lengths of the fish in the population at a particular moment are plotted as a frequency distribution there will be modes corresponding to sizes of fish produced at the peak of each spawning season, and growing at the average rate. Typically, with moderately long-lived fish the first one or two modes will be distinct, the following one or two can be detected with the eye of faith, or with more or less complex calculations which are not always much better, and the oldest age-groups are blurred together. The best approach to increase the number of age-groups for which the modal size can be estimated, i.e. the number of peaks that can be detected, is to estimate them in succession. The first one is usually well defined, particularly over its lower range; subtracting the estimate of the whole curve corresponding to the yougest age-group from the total distribution will enable the curve for the second youngest group to become more apparent; subtracting it may give the third, and so on, at least for one or two more groups than can be detected from the original data. This procedure is helped if some assumptions on the frequency distribution of the lengths of fish of a given age, e.g. that it is normal, which transformed by logarithms (i.e. plotting the logarithms of the numbers at each length against length) gives parabolas (e.g. Tanaka, 1956).

Given sampling at regular intervals this method can describe the growth in fair detail for the early period in the fishery, usually in more detail (or with less costs) than in obtainable from scale or otolith reading. Since this is often the period during which a knowledge of growth is most critical in assessing the stock, this approach can be very useful. However, it is valid only if the samples examined are truly representative of the population, and can be misleading if there is size selection. The danger of being seriously misled is not great when there is a single annual spawning, except at the size when the fish are just recruiting to the fishery. Then the early modes may only include the fastest growing fish and thus the progression of modes will underestimate the true growth. For species with an extended spawning season, e.g. tuna, the danger is much greater. Then the fishery may only sample a succession of spawning groups at approximately the same size, and the progression of modes, if any, may seriously underestimate the true growth rate. This seems to have been the case in the Pacific skipjack (Rothschild, 1967).

This method is usually successful for distinguishing the lengths corresponding to the two or three youngest year-classes in a fishery. For older fish the 12 months separation in age is not sufficient to prevent so much overlap between the length distribution of adjacent year-classes as to make it difficult or impossible to distinguish the corresponding modes, at least by eye. Methods have been proposed (e.g. Harding, 1949; Cassie, 1954; Bhattacharya, 1967) which can improve the objectivity of the determination of modes, and possibly enable one additional mode to be located. These and other problems of analysing length frequency data have been examined by McNew and Summerfelt (1978). MacDonald and Pitcher (1979) also give procedures for estimating the length frequencies of the component age-classes, in which additional information, for example on the ages of a small subsample, can be used to improve the statistical efficiency, and the reasonableness (in terms of the biological processes involved) of the estimates obtained.

The method can be extended to cover older groups under favourable circumstances, when the stock experiences large fluctuations in year-class strength, and it may be assumed that the rate of growth does not change much. Then the observed length composition of the population at a given season in a particular year (or of the population as sampled during that season by a given type of gear) may be compared with the average length composition at that season over a period of years. If the observations are expressed as deviations from the average, the presence of good or bad year-classes will appear as positive or negative anomalies. Since pairs of good year-classes may be separated by several years, these anomalies are less likely to lose their identity through the blurring together of adjacent modes (see exercise 4.1).

A method that can take advantage of some of the features of the Petersen method is the use of age–length keys. In its simplest form it is a special case of the normal statistical technique of stratified sampling. That is, the population being studied (e.g. the total landings from a particular fishery in a particular year) is stratified into length-classes, e.g. on the basis of extensive length measurements. The age-composition of each length group is then determined by samples (e.g. of otoliths) taken from each group. The number N_j of fish of age j in the population will then be given by the formula

$$N_j = \sum n_i p_{ij}$$

where n_i = number of fish in the ith length group, and p_{ij} = proportion of fish age j in ith length group, as estimated from the samples.

This procedure is unbiased and, if the age-composition of different size groups is very different, will be efficient. The efficiency will be increased by the right choice of the numbers to sample in each length-group. Usually the best strategy will be to concentrate sampling on the larger sizes, which will, because of slowing in growth rate and overlap between groups, contain more age-groups than the smaller size groups. In the extreme, if the smaller groups are shown to coincide with (Petersen) modes in the length composition it may be necessary only to sample one or two fish in these modes to check which age-group they are.

Like most useful tools age–length keys can be misused. Because of the variations in growth pattern between individual fish there will be considerable overlap in the length distributions of fish of different ages, i.e. a given length-group will contain several age-classes. The proportion of each age in a given length-group (which is the information needed for using an age–length key) will depend on the overall frequency of each age-class, i.e. on the original strengths of each year-class, and on the mortalities since recruitment. These will vary from year to year. It is particularly dangerous to use the same age–length key over a period in which mortality is changing. If, for example, it is increasing and a key is used from data collected early in the period, the proportion of older fish in each length-group will be progressively overestimated, leading to an underestimate of the increase in mortality. This bias in estimation of changes in mortality may be very great among the ages for which growth has slowed down, and in which each length-group contains several ages.

It has been pointed out by Pauly that it is possible to combine the two steps of splitting length–composition data to determine the mean length of age-classes and of fitting a growth curve. This approach is particularly useful when a series of length samples is available for a period of some years. Approximate modes are identified for each length sample, and the growth curve determined that gives the best overall fit to the length data. Pauly (Pauly and David, 1981) has developed computer programs ELEFAN 1 and ELEFAN 2 to handle the large amount of calculations involved. If these programs are applied without thought, especially to length data for which the modes are not clear, there is the danger (common to most thoughtless application of advanced techniques) that ridiculous results are obtained. The best procedure in practice is to analyse length data by simple graphical methods, and to attempt to fit growth curves by the methods of section 4.2.3. If this produces sensible results, then ELEFAN or similar programs can be used to extract the maximum amount of information from the data available.

4.2.3 Fitting growth curves

For population analysis it is desirable to express the growth of fish in a mathematical expression. The basic requirement is an expression which will give the size (in terms of length or weight) at any given age which agrees with the observed data of size at age, and which is in a mathematical form which can be incorporated reasonably easily in expressions for yield. Strictly, most population analysis is concerned more directly with growth rate, i.e. increase in weight or length per unit time, rather than with the size at various ages, because many problems in fishery assessment are essentially a matter of comparing weight gained by growth against that lost by natural mortality. Sometimes, for instance when considering the effect on an increase in size of first capture, it is particularly important to know the rate of growth over a comparatively short part of the total life span—that is, to know how long it will take a fish to grow from the original size of first capture to the new. There are therefore good reasons for preferring, other things being equal, a method of fitting equations to growth-rate data, rather than simply size at age.

Other desirable features of a growth equation are that the computational work involved in fitting it to observed data should be small, that the number of constants used should be few, that as far as possible these constants should have some biological meaning, and that if extrapolated to ages beyond those used in fitting the equation, the equation should not lead to unreasonable results.

Some of these considerations become less important when the calculation of, for example, the total weight of a catch of fish of various ages, can be done by computers, using observed empirical weight-at-age data, rather than by formal integration of a yield equation. Indeed, any computer with a moderate capacity can calculate expressions such as

$$Y = \sum_{r=1}^{n} N_r W_r$$

for quite large values of n, e.g. monthly values for a period of several years using values of N_r from a given expression (e.g. $N_r = N_0 e^{-rZ}$) and empirical, preset values of W_r with about as much ease as the explicit formulae derived from growth curves chosen to make calculations easy (e.g. exponential) could be done in the pre-computer era. In these circumstances there may not be any need, so far as simple stock assessments are concerned, to fit any mathematical growth curve at all. It may be sufficient to have values of weight-at-age to be provided as inputs to the computer calculations. However, a growth curve and especially the parameters derived from it (e.g. the values of W_∞ and K in the von Bertalanffy curve) provide a compact and handy way of describing the growth of a fish, and one that is particularly valuable in comparing growth in different fish stocks, or incorporating the effect of stock density or the environment on the fish population.

There exists a considerable literature on growth equations, which covers a wide range of possible equations, none of which seems to be entirely satisfactory in all situations. In fact it is most unlikely that a simple formula would always be able to describe the growth of even a single fish through most of its life, in which there could be greatly different conditions of food supply, reproductive strain, etc. These notes do not attempt to give a complete survey of growth equations, but will be mainly concerned with one particular equation, that ascribed to von Bertalanffy (1938), which satisfies the two most important criteria—it fits most of the observed data of fish growth, and can be incorporated readily into stock assessment models. Also, the parameters in this equation have some biological meaning (though the specific physiological derivation of the equation may not exactly be fulfilled in practice). Here, L_∞ or W_∞ is the maximum size that the animal would, on the average, reach if a fisherman did not catch it, or it did not fall to a predator or disease; K describes the rate at which it approaches this limiting size. It may therefore be considered in some way as a measure of the rate at which a fish lives, and therefore possibly related to other rates, particularly its mortality rate (see section 4.3.5). Roff (1980) points out that with modern computers the ability to incorporate growth equations into algebraic yield equations is no longer important. He puts forward a good case for the retirement of the von Bertalanffy equation, but it is likely for reasons of history and, to some extent, convenience, that it will continue to be commonly used.

If the length of a fish, crustacean, or lamellibranch is plotted against age, the result is usually a curve of which the slope continuously decreases with increasing age, and which approaches an upper asymptote parallel to the x-axis (see Figure 4.1). Curves of weight at age also approach an upper asymptote, but usually form an asymmetrical sigmoid, the inflexion occurring at a weight of about one-third of the asymptotic weight (see Figure 4.2).

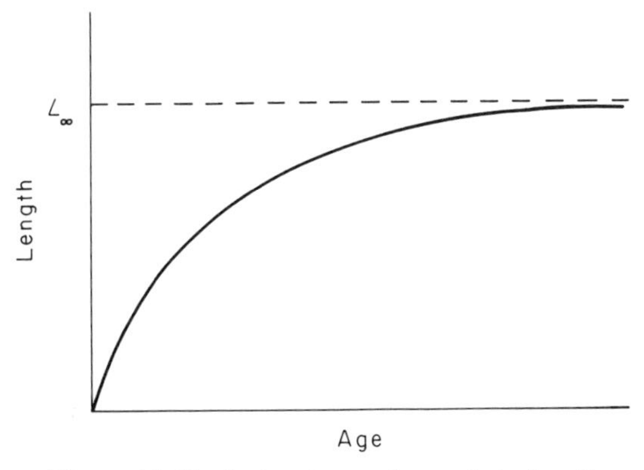

Figure 4.1 Typical pattern of growth in length

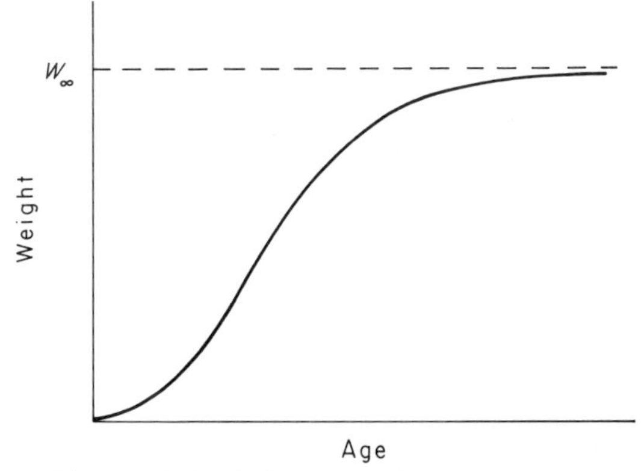

Figure 4.2 Typical pattern of growth in weight

If the rate of growth in length is plotted against length, the result is often well fitted by a straight line, cutting the x-axis at a point L_∞ beyond which the fish will not grow. This is of course the asymptote of the plot of length on age. If the rate of growth in length is linearly related to length, then in mathematical terms

$$\frac{dl}{dt} = K(L_\infty - l) \tag{4.1}$$

where L_∞ is the value of l for which the rate of growth is zero. This equation is in the general form of a straight-line relation, $y = ax + b$, with $a = -K$, $b = KL_\infty$. To integrate this differential equation, we write it in the form

$$\frac{dl}{L_\infty - l} = K\,dt$$

and therefore $-\log(L_\infty - l) = Kt + \text{constant}$, or $L_\infty - l = e^{-Kt} \times \text{constant}$, or $l = L_\infty - \text{constant} \times e^{-Kt}$. (4.2)

If the growth during the whole life of the fish exactly matched its growth during the period of observations from the data are taken, which will normally be over the commercial sizes, then at age $t = 0$, at the start of life, its size would be virtually zero. The constant would then be determined from equation (4.2) by putting $l = 0$ at $t = 0$, which gives a value of L_∞, i.e. the growth equation would be

$$l_t = L_\infty(1 - e^{-Kt})$$

Fish are seldom so well behaved; larval and juvenile fish have quite different feeding and other habits from commercial sizes, and therefore the growth curve of the latter sizes, if extrapolated backwards, seldom passes through the origin. Instead it will cut the x-axis at some age, t_0, the age at which the fish would have been zero size if it had grown according to the same pattern for all its life. Substituting in equation (4.2) we obtain a value of the constant represented in the form e^{Kt_0} which gives the growth curve

$$l_t = L_\infty[1 - e^{-K(t - t_0)}] \tag{4.3}$$

Equation (4.3) is identical with that deduced by von Bertalanffy (1938) on physiological grounds. He considered that growth in weight was the resultant of the difference between anabolic and katabolic factors, taken as proportional to surface area and weight respectively, i.e.

$$\frac{dw}{dt} = hs - kw$$

or, putting $w \propto l^3, s \propto l^2$, and simplifying

$$\frac{dl}{dt} = h' - k'l$$

where $h, h', k, k' =$ constants.

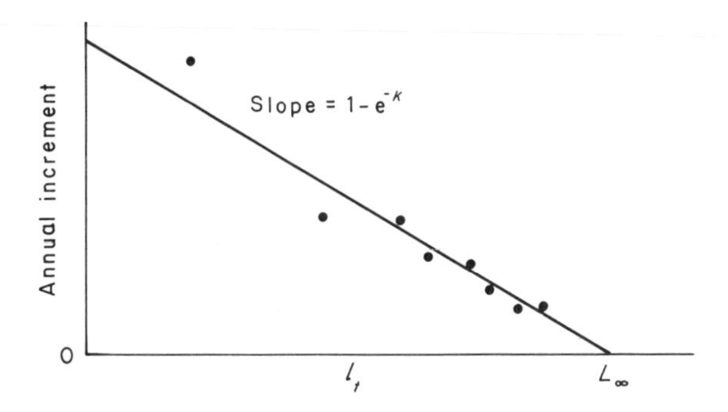

Figure 4.3 Estimation of growth parameters by plotting annual increment of length against initial length. Intercept on x-axis is L_∞ and slope $e^{-K} - 1$

In practice we seldom know the rate of growth. Instead we usually have the lengths of a fish at certain instants of time, or more usually the average lengths of a group of fish (e.g. a year-class) during a certain interval of time (e.g. during a particular fishing season). From these data we can measure the increments, e.g. between successive seasons. Following the implications of equation (4.1) we can plot these increments against length, say the length at the beginning of the interval. This will give a line with a negative slope. It will cut the intercept (subject to the usual problems of sampling and other errors) at L_∞, but the slope will not quite be equal to K. Mathematically, we have for data at equal time intervals of length T, from equation (4.3)

$$l_t = L_\infty[1 - e^{-K(t-t_0)}]$$

$$l_{t+T} = L_\infty[1 - e^{-K(t+T-t_0)}]$$

$$l_{t+T} - l_t = L_\infty e^{-K(t-t_0)}(1 - e^{-KT})$$

$$l_{t+T} - l_t = (L_\infty - l_t)(1 - e^{-KT}) \tag{4.4}$$

The plot of increment, $l_{t+T} - l_t$, against initial length, l_t, therefore gives a line, slope $-(1 - e^{-KT})$, and an intercept, on the x-axis, of L_∞ (see Figure 4.3). The important special curve is $T = 1$ year, when the slope $= -(1 - e^{-K})$. An alternative form of equation (4.4) is

$$l_{t+T} = L_\infty(1 - e^{-KT}) + l_t e^{-KT} \tag{4.5}$$

which for $T = 1$, is the well-known Ford–Walford plot, of l_{t+1} against l_t, which gives a straight line, slope e^{-K}, and an intercept on the 45° line, where $l_t = l_{t+1}$, of L_∞ (see Figure 4.4). This plot is essentially the same as the plot of increment $(l_{t+1} - l_t)$ against initial length. The points will seem to fit the Ford–Walford line better, but because the intersection of the regression line and the 45° line is very oblique, L_∞ is in fact estimated with equal precision by the two plots, and if the fitting is done graphically rather than by calculating regression lines, e.g. by least squares, greater errors in drawing are likely to be introduced in the plot of l_{t+1} against l_t.

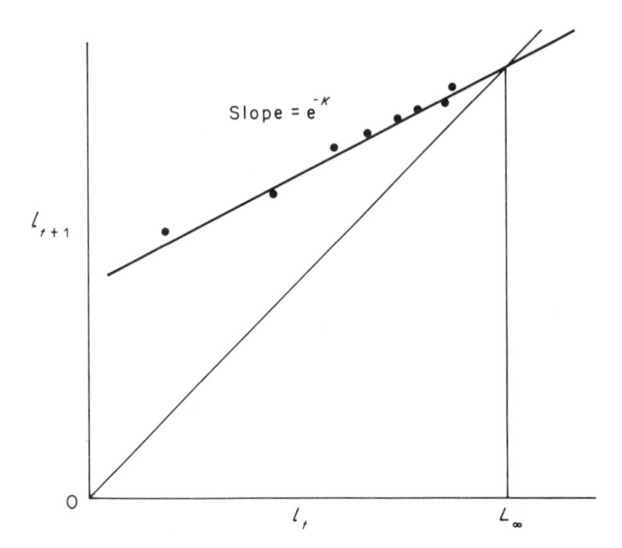

Figure 4.4 Estimation of growth parameters by plotting length at a given age against the length a year earlier (the Ford–Walford plot). Intercept on 45° line is L_∞ and slope is e^{-K}

Here, L_∞ and K can be determined directly from the lines fitted to these plots either by regression analysis, or by eye; then t_0 can be estimated from equation (4.3) for any particular observation of length at age. For this, equation (4.3) is best rewritten as

$$e^{-K(t-t_0)} = \frac{L_\infty - l_t}{L_\infty}$$

or

$$t_0 = t + \frac{1}{K}\log_e \frac{(L_\infty - l_t)}{(L_\infty)} \tag{4.6}$$

While an estimate of t_0 can be thus obtained for each age for which the mean length is known, these estimates will not be equally good. Those from old fish will be highly variable, because a small difference in l_t makes a big difference to the estimate of t_0 when l_t is nearly equal to L_∞, while the mean length of the youngest fish may be biased because only the bigger fish of these ages appear in the catches. The best estimate of t_0 would appear to be the mean of the estimates of t_0 obtained from the younger, but fully recruited, age-groups. Where good computing facilities are available, other methods of fitting can be used (e.g. Bayley, 1977) which are statistically more efficient. They do not differ greatly in principle from the methods of graphical plotting and simple linear regression described above. A graphical plot, e.g. of $(l_{t+1} - l_t)$ against l_t is in any case always recommended to give a visual check of how well the data fit the model.

Alternatively, only L_∞ is estimated from the plots of increment against initial length. From this $\log_e[(L_\infty - l_t)/L_\infty]$ can be calculated, and plotted against t. From equation (4.6) this should give a straight line, slope $-K$ intercept on the t-axis equal to t_0.

Data on length at annual intervals can be obtained and tabulated in various ways. Probably the most straightforward is the analysis of the growth of a single fish—its length at the end of each year of life being estimated by measuring the position of each annual ring along an axis of a scale, otolith, or other hard structure. Alternatively, the growth of a single year-class can be followed over a period of years—i.e. for the 1956 year-class the increment in the fifth year of life is the difference between the mean length of five-year-old fish at the end of 1961 and four-year-old fish at the end of 1960. Both these methods are concerned with the growth of a group of fish during their life, the growth during each year of life providing one point in the plot. Thus the growth increments refer to different calendar years; for example, for the 1956 year-class the increment in the third year of life is put on in 1959, possibly under conditions very different to those in 1961, the fifth year of life. Growth in a particular year can be studied by taking the growth of different year-classes—e.g. data of growth in 1960 are given by the growth of the 1956 year-class in its fifth year of life and of the 1957 year-class in its fourth year of life—each year-class provides one point in the plot. This is a particularly useful method for studying the effect of the environment—food, temperature, density, etc. on growth. Finally, growth can be estimated from a single year's data, for example, from data in 1960 the increment in the fifth year of life can be estimated from the difference in lengths of the 1955 and 1956 year-classes, the fourth year's increment from the difference between the 1956 and 1957 year-classes, etc. These differences do not, in fact, correspond to the growth of any particular fish or group of fish, and the method should be used only when other methods cannot. An improvement is to take the average size at each age over a number of years, but in this case it is usually possible, and more satisfactory, to analyse the growth of the individual year-classes or in each calendar year.

The basic von Bertalanffy equation (4.3) can be manipulated into various forms suitable for different purposes. For example, following Allen (1976) a form suitable for comparing growth rates from different stocks of the same species, is obtained by assuming that the value of K is the same for all stocks, and is known. Then writing $e^{-K} = r$, equation (4.3) can be written as

$$l_i(t) = a_i + b_i r^t$$

where l_i, a_i, b_i = values for the ith stock. Regression of $l_i(t)$ on r^t then gives values of $a_i(L_\infty)$ and $b_i(= L_\infty e^{Kt_0})$ which can be compared between stocks. The use of a single parameter $\omega(= KL_\infty)$ has been suggested by Gallucci and Quinn (1979) to avoid the problems associated by the close correlation between estimates of K and L_∞. It is useful in distinguishing differences in the early growth rate of different populations, but seems less sensitive when the differences occur mainly in the larger fish.

4.2.4 Length and weight

Traditionally most estimates of growth curves have been made, as in section (4.2.3), in terms of length rather than weight. This procedure has justification in that most original observations are in terms of length, and that the calculations tend to be easier. However, the importance of a fish, either in the ecosystem (e.g. as consumer, or as a source of food to predators) or to the fisherman, is better measured by its weight. The calculation of the relation between length and weight is easily done, and though not a major research finding, often supplies a topic for a minor scientific paper in the early stages of an investigation before more significant work can be completed. Since the weight is usually proportional to some power (usually the cube) of the length, the estimation is best done by expressing the relation $w \propto l^b$, i.e. as

$$\log w = a + b \log l$$

and calculating the regression of log w on log l by the standard statistical methods. This will usually give a value of b, that is close to, but not quite equal to 3. Expressing weight as proportional to the cube of length gives a more convenient and compact formula, and it is therefore useful to test the significance of any difference of the least-squares estimate of b from 3. If there is no significant difference, we can write $w = kl^3$, where k can be defined as the condition factor (Fulton, 1911) and can be estimated as the mean for all fish sampled. The condition factor provides a convenient measure for comparing the weights of fish from different areas or in different seasons. If the length–weight relation is approximately a cubic, then writing W_∞ as the limiting weight ($= kL_\infty^3$), the von Bertalanffy equation (4.5) can be written in the form

$$W_t = W_\infty [1 - e^{-K(t - t_0)}]^3$$

Alternatively, we can note that modern computing facilities do not require the use of a single weight-at-age formula valid over the whole life span of a fish in the fishery, and that for short intervals an exponential growth curve is both easier to fit and easier to incorporate into calculations. That is, the weight can be written as

$$W_{r+t} = W_r e^{G_r t} \quad (4.7)$$

where W_r denotes the weight at the beginning of the rth growth interval, and G_r the growth rate during that interval.

When considering the economic returns to fisheries, the value of the catch is important. While, under given market conditions, this is nearly proportional to the weight, the price per unit weight is seldom exactly the same for all sizes of fish. Usually the bigger fish are more valuable than small fish. It then may be desirable to calculate a growth curve in terms of the value of the individual fish

$$V_{r+t} = V_r e^{G_r t} \quad (4.8)$$

When the original data are in terms of weight, and it is desired to fit the von Bertalanffy equation, the easiest procedure is to take the cube root of the weight as an index of length; fit the equation to the resultant 'length' data, and cube the result to give the equation for weight.

The seasonal variations in value, and the systematic seasonal departures of weight, and to a lesser extent length, from a smooth growth curve fitted to mean annual values, suggest that some modification to that curve is desirable. This may be done by incorporating modifications of the theoretical equation (e.g. Cloern and Nichols, 1978; Pauly and Gaschütz, 1979). In practice it may be easier to incorporate observed weight (or values) at age date into empirical computer calculations.

A more general growth equation has been given by Richards (1959); with a slight change from his notation this is

$$w_t^{1-m} = W_\infty^{1-m}[1 - ae^{-kt}] \tag{4.9}$$

For various values of m this equation becomes one or other of the common growth equations. Thus, if $m = \frac{2}{3}$, equation (4.9) becomes the same as equation (4.3) for the von Bertalanffy curve. If $m = 2$ the equation, with some rearrangement, becomes the autocatalytic equation

$$w_t = \frac{W_\infty}{1 + be^{-Kt}}$$

and for $m = 0$, the monomolecular equation

$$w_t = W_\infty(1 - ae^{-Kt})$$

and it can be shown that in the limit as $m \to 1$, the equation becomes the Gompertz $\log w_t = \log W_\infty(1 - ae^{-Kt})$. This curve satisfies a relation similar to that of equation (4.1), only in terms of log length rather than length, i.e.

$$\frac{d\log l}{dt} = K'(\log L_\infty - \log l)$$

4.2.5 Conversion of length to age

The expressions above give the weight or length of a fish in terms of its age. Sometimes the inverse procedure is required, i.e. it is desired to know the age of fish of a given length, e.g. selection data are normally in terms of length, but for incorporation in yield equations need to expressed in terms of age.

If the seasonal growth pattern is very marked, and all fish of a particular age are nearly the same length, the actual mean age of fish of a given length may be quite different from the mean age determined from the average annual growth curve, particularly at the beginning and end of the season of rapid growth. If this is the case, the conversion of length to age is best made empirically from the observed age–length curve. Also, particularly if there is much individual variation in growth, the curve of mean length at a given age will be different from the curve of mean age at a given length (just as generally the regression line of y on x is different from the regression line of x on y). Usually, however, it is sufficient to convert length to age from a growth equation fitted to all the observed data of mean length at age, e.g. the von Bertalanffy equation

$$l_t = L_\infty[1 - e^{-K(t - t_0)}]$$

To obtain t in terms of l we divide both sides by L_∞, and subtract from unity, giving

$$\frac{L_\infty - l_t}{L_\infty} = e^{-K(t-t_0)}$$

Taking natural logs of both sides gives

$$\log_e \frac{L_\infty - l_t}{L_\infty} = -K(t - t_0)$$

and therefore

$$t = \frac{1}{K} \log_e \frac{L_\infty}{L_\infty - l_t} + t_0 \tag{4.10}$$

4.3 MORTALITIES

4.3.1 General concepts

Each year a proportion of the fish alive at the beginning of the year will die—some by predation, disease, or other natural causes, and some by being caught—while others will survive until the beginning of the next year. Mathematically, we have

$$N_1 = P + D + O + C + N_2$$

where N_1, N_2 = the numbers at the beginning of the first and second year, P, D, O = numbers dying of predation, disease, and other causes, and C = numbers caught. The most obvious way of expressing these deaths is as proportions of the numbers at the beginning of the year, that is, writing

$$\text{annual predation rate} = P/N_1$$

$$\text{annual rate of death by disease} = D/N_1$$

$$\text{annual rate of death by other causes} = O/N_1$$

$$\text{annual rate of death by exploitation} = C/N_1 = u$$

$$\text{annual death rate by all causes} = 1/N(P + D + O + C) = A$$

$$\text{annual survival rate} = N_2/N_1 = S = 1 - A$$

While these expressions are obvious and easily understood, they are in practice not very useful in studying the dynamics of the population, since they lead to clumsy algebraic expressions when the effects of different causes of mortality are combined, or when considering the effects on the annual rate of exploitation of changes in the amount of fishing. For example, doubling the amount of fishing (e.g. the number of trawlers) will not double the number of fish caught, because the additional ships will reduce the catch of those already fishing.

It is therefore better to consider the instantaneous rates, i.e. the rates applying over a short period of time, dt, during which the numbers in the population do not change significantly, so that the numbers dying from any one cause are not affected by the numbers dying from any other cause. Then, combining all causes of natural (non-fishing) mortality, these deaths will be proportional to the instantaneous natural mortality coefficient M, and the numbers caught to the instantaneous fishing mortality coefficient F, so that

$$(\mathrm{d}P + \mathrm{d}D + \mathrm{d}O) = MN_t\mathrm{d}t \tag{4.11}$$

$$\mathrm{d}C = FN_t\mathrm{d}t \tag{4.12}$$

and total deaths, which are equal to the decrease in the population numbers, can be written as proportional to the total mortality coefficient Z

$$-\mathrm{d}N = ZN_t\mathrm{d}t \tag{4.13}$$

From these equations it is clear that the total mortality coefficient is the sum of all the other coefficients

$$Z = F + M \tag{4.14}$$

and further that the fishing mortality coefficient is proportional to the amount of fishing, or fishing effort f, i.e.

$$F = qf \tag{4.15}$$

It should be noted that equation (4.15) is not an assumption that needs to be justified in any given situation, but follows directly from the nature of fishing effort and the fishing mortality. The assumption that has to be made in relation to any particular fishery is that the data on nominal effort (number of vessels, number of purse-seine sets, etc.) which are available for that fishery are indeed reliable measures of the true effort. The distinction is more than a semantic one because it does help to concentrate attention where it is needed—on the operational behaviour of the fishing fleet (see section 2.3).

Writing equation (4.13) in the form

$$\frac{1}{N_t}\mathrm{d}N = -Z\mathrm{d}t$$

and integrating, we obtain

$$\log N_t = -Zt + \text{constant}$$

or

$$N_t = N_0\mathrm{e}^{-Zt} \tag{4.16}$$

where N_0 = numbers alive at time $t = 0$.

Equations (4.14), (4.15), and (4.16) are among the most basic equations of fish population dynamics. From them, expressions for the annual rates described earlier can be obtained. From (4.16) we have $N_2 = \mathrm{e}^{-Z}N_1$, and hence

$$S = \text{survival rate} = \frac{N_2}{N_1} = \mathrm{e}^{-Z}$$

$$A = \text{annual death rate} = 1 - \mathrm{e}^{-Z}$$

Of these deaths, proportions F/Z, M/Z will be due to fishing and natural mortality respectively, so that

$$\text{annual rate of exploitation} = u = \frac{F}{Z}(1 - e^{-Z})$$

$$\text{and annual rate of natural deaths} = \frac{M}{Z}(1 - e^{-Z})$$

So far the mortality coefficients have been treated as constant. In fact they will vary. Predation is likely to be greater on small fish, while older fish may be more subject to disease, and though there is little evidence of fish dying of old age, there is often a high post-spawning mortality (complete in the case of Pacific salmons). Fishing mortality will vary from year to year, and also few gears are so unselective that there is not some variation in mortality with age of fish.

Then we should write

$$Z_t = F_t + M_t$$

where the subscript denotes the values at time t, and

$$\frac{1}{N_t} = -Z_t \mathrm{d}t$$

$$\log N_t = \int Z \mathrm{d}t$$

and equation (4.16) becomes

$$N_t = N_0 \exp\left(-\int_0^t Z_t \mathrm{d}t\right) \tag{4.17}$$

The following sections deal with the problems of estimating F, M, and Z, first in the case of constant values and then considering possible variations with time and age of fish.

4.3.2 Estimation of total mortality

In the ideal world estimation of total mortality is easy. All that is required is a pair of estimates, n_1, n_2 of the numbers of a certain group of fish at two points in time t_1, t_2. Then during the time interval the numbers will fall by a proportion $e^{-Z(t_2-t_1)}$ and therefore we can estimate Z from the relation

$$n_2/n_1 = e^{-Z(t_2-t_1)}$$

or

$$Z = -\frac{1}{t_2 - t_1}\ln(n_2/n_1) \tag{4.18}$$

or, if the observations are made one year apart, so $t_2 - t_1 = 1$,

$$Z = -\ln(n_2/n_1). \tag{4.19}$$

Given the basic equation (4.19) for a single group of fish, observed a year apart, it is easy to see how it can be expanded to provide estimates from larger sets of data. For example, if n_1, n_2 refer to the 1966 year-class observed when 5 and 6 years old at the beginning of 1971 and 1972, and we also have observations of the abundance of fish age 7, 8...12 in each year, then there will be a set of estimates from the 1965 and earlier year-classes, equal to (with the obvious notation) $\ln {}_2n_7/{}_1n_6$, $\ln {}_2n_8/{}_1n_7$, etc. There are various ways of combining these to give a single estimate of mortality. A common and convenient one is to consider all the fish older than a given age in the first year, say five, writing

$$S = \frac{{}_2n_6 + {}_2n_7 + \dots}{{}_1n_5 + {}_1n_6 + \dots}$$

and

$$Z = -\ln\left(\frac{{}_2n_6 + {}_2n_7 + \dots}{{}_1n_5 + {}_1n_6 + \dots}\right)$$

This form is likely to be a good method of obtaining a single estimate from a set of age data since it gives most weight to the more abundant age-groups which are likely to be the best estimated.

Alternatively, especially if data are available from a series of years, it is possible to construct a table of estimates of $Z = {}_xZ_t$, for different years x and ages t; from this the overall mean value of Z can be calculated and also, using an analysis of variance, or by simple examination of the mean values for different ages and years, to see whether there are any systematic changes in mortality.

In practice matters are not as simple as this; the main difficulties are:

(a) numbers are usually observed as averages over a period of time, rather than at a particular moment;
(b) observations are usually indices of abundance (e.g. catch per unit effort, c.p.u.e.) rather than absolute numbers, and the relation of the index to actual numbers may change;
(c) it may be difficult to be sure that the same group of fish is being observed at the two moments. Usually this is dealt with by considering groups of fish for which the ages of fish are known, e.g. the 1977 year-class.

The average numbers $\bar{N}$ of fish alive during some period length T will be given by

$$\bar{N} = \frac{1}{T}\int_0^T N_t \mathrm{d}t = \frac{1}{T}\int_0^T N_0 \mathrm{e}^{-Zt} \mathrm{d}t = \frac{N_0}{ZT}(1 - \mathrm{e}^{-ZT})$$

which for $T = 1$ reduces to

$$\bar{N} = \frac{N_0}{Z}(1 - \mathrm{e}^{-Z})$$

If we denote by appropriate suffices the numbers and mortalities during successive annual periods 0 and 1, we have

$$\bar{N}_0 = \frac{N_0}{Z_0}(1 - e^{-Z_0})$$

$$\bar{N}_1 = \frac{N_1}{Z_1}(1 - e^{-Z_1})$$

and further, $N_1 = N_0 e^{-Z_0}$.

Following the obvious procedure suggested by equation (4.19), we can estimate an 'average' mortality Z' during the two years from the ratio of the average numbers in the two years, as

$$Z' = \ln(\bar{N}_1/\bar{N}_0)$$

In fact

$$\bar{N}_1/\bar{N}_0 = e^{-Z_0}\left(\frac{1 - e^{-Z_0}}{Z_0}\right)\left(\frac{Z_1}{1 - e^{-Z_1}}\right)$$

or

$$\ln(\bar{N}_1/\bar{N}_0) = -Z_0 + \ln\left(\frac{1 - e^{-Z_0}}{Z_0}\right)\left(\frac{Z_1}{1 - e^{-Z_1}}\right) \quad (4.20)$$

That is, the logarithm of the ratio of the *average* abundance will be equal to the mortality in the first year, with the addition of a correction term. Examination of the last term in equation (4.20) shows that if $Z_0 = Z_1$ it reduces to zero. That is, as might be guessed intuitively, if the mortality is constant the ratio of average abundance gives the true mortality. Further, if both Z_0 and Z_1 are small, the two parts of the last term in (4.20) are nearly equal to 1, and again the mortality is correctly estimated.

Further, if the mortality rates in the two years are written in terms of the average mortality $\bar{Z}$, and the increase dZ between the two years, i.e.

$$Z_0 = \bar{Z} - \tfrac{1}{2}dZ$$
$$Z_1 = \bar{Z} + \tfrac{1}{2}dZ$$

then equation (4.20) can be written as

$$\ln \bar{N}_1/\bar{N}_0 = -\bar{Z} + \tfrac{1}{2}dZ + \text{an expression containing } \bar{Z} \text{ and } dZ$$

If the right-hand side is expanded in powers of dZ, the coefficients of powers of dZ cancel out and it will reduce to

$$\ln \bar{N}_1/\bar{N}_0 = -\bar{Z} + \text{terms in } (dZ)^2$$

i.e. to a close approximation, which becomes very close if dZ is small, the ratio of average abundance provides a measure of the average mortality over the two years. This will clearly also hold true, to a reasonable approximation, where the mortality rate changes continually during each of the two years and not merely as an abrupt change from one year to the next. In the following sections the discussions will usually (except for cohort analysis) be expressed in terms of point estimates of abundance, though the commonest applications will be in terms of the average over a period. The distinction should be kept in mind, though the practical difference is often small.

When the actual abundance is not known, some index, usually the c.p.u.e., may be available. That is, for numbers at some points a year apart, and examining the simplest situation

$$n_1 = \text{c.p.u.e.} = q_1 N_1$$
$$n_2 = q_2 N_2 = q_2 N_1 e^{-Z}$$

The obvious approximation to the mortality rate, Z, is $-\ln(n_2/n_1)$ which given by the equation

$$-\ln n_2/n_1 = -\ln(q_2 N_2/q_1 N_1) = -\ln q_2/q_1 - \ln(N_2/N_1)$$
$$= -\ln q_2/q_1 + Z$$

That is, there is an extra term, equal to the logarithm of the ratio of the catchability coefficients q; only if these are equal can the catch per unit data be used directly to estimate total mortality rate. This is a severe restriction since the catchability coefficient is seldom precisely constant (see section 2.3) (i.e. c.p.u.e. does not measure abundance consistently), and is likely to vary with the age of fish ('selectivity' of the gear in a wide sense), and with time, due to, for example, improvements in the efficiency of the gear.

4.3.3 Catch curves

One way of dealing with the latter aspect is to consider data collected at one instant of time. In a sample taken at one particular moment the ratio of different age-groups will be unaffected by changes in catchability with time. As each successive age-group will have been exposed to one more year's mortality than the next younger group, their ratio will be approximately equal to the annual survival. This suggests that an estimate of total mortality rate which is free from some source of error could be obtained from a single age-sample.

Mathematically, at the beginning of the xth year, the abundance ${}_xN_t$ of fish age $t + r$, where r is the age at recruitment, will be given by

$${}_xN_t = R_{x-t} \exp - \left(\sum_1^t Z_{x-i} \right) \tag{4.21}$$

where R_{x-t} = number of age r recruits at the beginning of year $x - t$, and Z_{x-t} = total mortality in year t.

Similarly,

$${}_xN_{t+1} = R_{x-t-1} \exp - \left(\sum^{t+1} Z_{x-i} \right)$$

Therefore

$${}_xN_{t+1}/{}_xN_t = - \frac{R_{x-t-1}}{R_{x-t}} \exp(-Z_{x-t-1}) \tag{4.22}$$

Also, bearing in mind that the catchability coefficient can vary from year to year, and can depend on the age of fish, the catch per unit of fish age $t+r$ in year x, ${}_x n_t$, can be written

$${}_x n_t = {}_x q_t \, {}_x N_t$$

and

$${}_x n_{t+1} = {}_x q_{t+1} \, {}_x N_{t+1}$$

so that the ratio of the logarithms of the catches per unit effort will be equal to

$$\ln({}_x n_t / {}_x n_{t+1}) = \frac{{}_x q_t \, {}_x N_t}{{}_x q_{t+1} \, {}_x N_{t+1}}$$

or, if equation 4.22 holds,

$$\ln({}_x n_t / {}_x n_{t+1}) = \ln({}_x q_t / {}_x q_{t+1}) + \ln(R_{x-t-1}/R_{x-t}) + Z_{x-t-1} \tag{4.23}$$

That is, the ratio of the catches per unit effort of successive age-groups in a particular year will provide an estimate of the mortality in the year before the younger year-class recruited to the fishery (since in later years the two groups of fish suffered the same mortalities). In addition there are correction terms to take account of any difference in age-specific catchability ('selection'), and in the initial numbers of recruits into the two year-classes. It may be noted that strictly if catchability varies with age, the mortality experienced by the two year-classes will not be exactly the same in any year, but for the present purposes the difference can be ignored.

While equation (4.23) can be used for arithmetic calculations to estimate mortality from any pair of age-groups, the best application of this approach is a graphical one, combining all the data from all age-groups. To do this we note that

$$\ln({}_x n_t) = \ln({}_x q_t) + \ln({}_x N_t)$$

and hence from equation (4.21), assuming the mortality rates Z_{x-t} are constant $= Z$

$$\ln({}_x n_t) = \ln({}_x q_t) - \ln(R_{x-t}) - Zt \tag{4.24}$$

That is, the logarithm of the numbers caught per unit effort should be a linear function of age; and if a linear regression is fitted, the slope of the line provides an estimate of Z. The fitting can be done by normal statistical regression methods, but is more usually done graphically—the so-called catch curve (Ricker, 1975)—which is very easy by plotting directly on semi-log paper (see Fig. 4.5). There are two reasons for this: first, the common advantages of the normal procedures of statistical fitting—estimates of confidence limits—do not immediately apply because the variance about the regression is likely to increase rapidly with increasing age; second, bearing in mind that the catch curve can be thought of giving a fossil record of mortality rates during the period since the

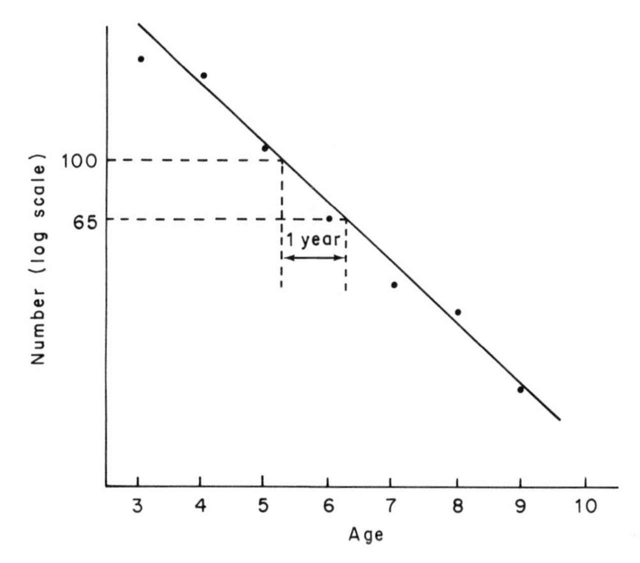

Figure 4.5 Estimation of total mortality from a catch curve. The slope is $-e^{-z}$. The survival s ($=e^{-z}$) can be easily read from the curve by picking a point where the number caught is convenient, e.g. 100, and reading off the numbers caught one year later

oldest fish present recruited into the fishery, visual examination of scatter of points about the line can give insight into possible changes in mortality. For example, a steeper slope in the part corresponding to more recent years, i.e. for younger fish, is an indication of increased mortality.

The interpretation of catch curves must be done with caution because trends in the other factors in equation (4.24) can produce similar effects. For example, if recruitment has recently been increasing, the result will also be an apparent steepening of the left-hand part of the catch curve, and a general trend of increasing recruitment over the whole period cannot be distinguished from a slightly higher mortality rate.

Changes in catchability with age cause special difficulties in interpretation. If, for example, it decreases, then, other things being equal, the c.p.u.e. will be lower but the decreased catchability (and therefore fishing mortality) will result in there being more old fish than there otherwise would be, and the observed c.p.u.e. may be very close to that expected on the basis of constant catchability (cf. Ricker, 1975, Figure 2.4, curve A). Thus, although when the catch curve deviates from a straight line this is good evidence that one or both of recruitment or mortality has not been constant, the fact that the catch curve does not, so far as can be seen from the data, deviate from a straight line, does not necessarily indicate that catchability (and hence fishing mortality) is indeed the same for all post-recruit age-groups.

4.3.4 Use of length data

The size composition of a population is clearly related to the total mortality; the lower the mortality, the more old, and therefore large, fish there will be. There are various ways, depending on the amount that is known about the growth pattern of the fish, in which this general relation can be used to obtain information on the total mortality in cases where age-composition data are not available.

The simplest approach is to note that the mean length l of fish greater than any particular length l_c is given by

$$\bar{l} = l_c + \frac{K}{Z+K}(L_\infty - l_c)$$

from which the expression for total mortality

$$Z = \frac{K(L_\infty - \bar{l})}{\bar{l} - l_c}$$

can be deduced.

Even when regular age-data are not available, sufficient may be known about the growth of the fish, e.g. from tagging, to provide estimates of K and L_∞. Alternatively, the size of the largest fish occurring in the population may be used as a first approximation to L_∞. In this case estimates of Z/K rather than Z itself are obtained; these are still useful, since a value of M/K can often be assumed, from which the ratio of F and M can be derived.

A slightly different expression has been obtained by Ssentongo and Larkin (1973). Writing

$$y = \log(1 - l/L_\infty) \quad \text{and} \quad y_c = \log\left(1 - \frac{l_c}{L_\infty}\right)$$

they derived an unbiased estimate from a sample of n fish as

$$Z/K = \frac{n}{n+1} \cdot \frac{1}{\bar{y} - y_c}$$

which for large n reduces to

$$Z/K = \frac{1}{\bar{y} - y_c}$$

(It must be noted that $\bar{y}$ has to be calculated from the individual values of y, and is not equal to $\log(1 - \bar{l}/L_\infty)$.)

Length-frequency data can be used in a similar way as age-frequency data to produce 'catch curves'. If the logarithm of the numbers in each length group are plotted against length, the result will look like a typical catch curve, and the slope of the right-hand side over any range in which the growth in length is approximately linear will be proportional to the total mortality. At larger sizes it takes longer for a fish to grow through a given length, so each size interval, e.g. a 5 cm length group, will include an increasing number of age-groups. This effect can be corrected for by writing

$$n'_x = \frac{n_x}{t_x}$$

where n_x is the number in the xth length group (between lengths l_x and l_{x+1}), and t_x is the average time it takes a fish to grow from l_x to l_{x+1}. This latter can be derived from a growth curve in which age is expressed as a function of length—which will be slightly flatter than the usual curve in which length is expressed as a function of age.

If then $\log n'_x$ is plotted against x the result should, if mortality is constant, be a straight line whose slope is proportional to Z.

4.3.5 Cohort analysis

Instead of considering the numbers caught of different year-classes in the same year, it is possible to consider the catches of the same year-class, or cohort, of fish in successive years. This obviously removes problems of differences in the original strengths of year-classes, but also, with certain assumptions, can produce estimates of mortality rates independent of the catchability coefficient, or data on fishing effort.

To do this we note that the catch C_t of fish age t is given by

$$C_t = N_t \frac{F_t}{F_t + M}(1 - e^{-(F_t + M)}) \qquad (4.25)$$

where N_t = population at the beginning of the year, and

$$N_{t+1} = N_t e^{-(F_t + M)}. \qquad (4.26)$$

After substituting for N_t in equation (4.25), it will be seen that, if we are dealing with n age-groups in the catches, we will have n equations corresponding to (4.25) for each of n successive years during which the year-class or cohort appears in the catches. In these n equations there will be $n + 2$ unknown quantities (n values of F, M, and one value of N). The equations therefore cannot be solved unless we make some assumptions to reduce the number of unknown quantities. The assumptions usually made are of the value of natural mortality, and of the fishing mortality on the oldest age. As regards the latter assumption it can be shown (e.g. Pope, 1972; Agger *et al.*, 1971) that for a long-lived fish in which fishing mortality is the main cause of death, differences in the assumed fishing mortality in the oldest age-group have progressively smaller effects on the estimates for the younger and more important age-groups. The assumption made about natural mortality can, however, be critical.

This approach (sometimes called virtual population analysis because of the history of its development in the North Atlantic (Jones, 1964; Gulland, 1965), which proceeded independently from the development in the Pacific (Murphy, (1965)) has proved of very wide application in many commercial fisheries. The features of a fishery that make this approach valuable are the existence of a large number of age-groups in the fishery (so that the loss of one year's data in assuming the value of F on the oldest group is not important); a long series of age-composition data; and a complex and variable fishery (so that F is likely to vary with both age and year). These conditions are fulfilled by many of the most important fisheries of the North Atlantic.

The arithmetic involved in the estimation when calculating back from an assumed value of F on the oldest age-group can become complicated, because the value of F on the younger fish appears, in a non-linear form, in all the expressions of catches, population numbers, etc.

One procedure is to obtain two expressions for r_n, the catch in year n, expressed as a proportion of the numbers alive at the *end* of the year (this proportion may be greater than unity). In terms of the fishing mortality during the year, we have

$$r_n = C_n/N_{n+1} = \frac{F_n}{F_n + M}(1 - e^{-(F_n+M)}) \Big/ e^{-(F_n+M)} \tag{4.27}$$

The right-hand side is a simple expression, containing only F_n and M. For a given value of M it can readily be calculated and tabulated as a function of F_n. Also, the left-hand expression for r contains only the catch in year n (which is known), and the value of N_{n+1}, which will be known once the value of F for year $n+1$ is known. That is, once F_{n+1} is known, r_n can be calculated, and the corresponding value of F_n read off from the table.

The basic procedure is simple, but laborious if repeated for several cohorts (year-classes), and also, as it should be, for several assumptions about the value of F on the oldest age-group and about M. It is therefore well suited to computer handling, and a number of programs exist (e.g. Schumacher, 1971).

Though this procedure allows estimates of F to be made for each age in each year-class quite separately, the accuracy of the final estimates for a fishery can be increased by noting that different cohorts are likely to experience very similar changes in fishing mortality from year to year (due to changes in the total amount of fishing), and for different ages (due to the 'selectivity' of the fishery. That is denoting ${}_tF_n$ as the fishing mortality on fish age t during year n, we can write

$${}_tF_n = \bar{F} + g_n + {}_tS + {}_ta_n \tag{4.28}$$

where $\bar{F}$ average value of F for all ages and years, and g_n, ${}_tS$ = the effects of the general level of fishing effort and of selectivity respectively, and a_n can be considered as an 'error' or interaction term.

Equation (4.28) can be used, given a suitable array of estimates of ${}_tF_n$, as the basis of an ordinary analysis of variance. Estimates of the set of values of f_n, for all years, and ${}_tS$, for all ages, can then be obtained. If it can be considered that there is no significant interaction (i.e. 'selectivity' does not vary from year to year), and the values of ${}_ta_n$ represent an essentially random element, then the best estimates of the fishing mortality on each age and year will be

$$_t\hat{F}_n = \bar{F} + \hat{f}_n + {}_t\hat{S} \tag{4.29}$$

The decision on whether or not to accept the a's as error terms should be based to some extent on an examination of the values obtained, and to a much greater extent on a practical experience of the fishery and of year-to-year changes in its operations. If there are mainly changes in the number of vessels, then the effective selectivity probably does not change. On the other hand, if there are changes in the type of gear used (or in the proportions of vessels using different types of gear), or in the grounds or seasons fished (especially if there are noticeable differences in the sizes of fish caught on different grounds or at different seasons), then it would be unwise to assume a constant pattern of 'selectivity' from year to year.

Alternatively, instead of examining the values of F obtained from the analyses of individual cohorts, the model suggested by equation (4.28) can be applied to the original catch-at-age data to obtain estimates that give the best description of the data as a whole. This involves more complex computations than can be described here, and the reader is referred to the original papers describing the use of this approach, such as Pope (1977), Doubleday (1976), and Gray (1977).

Another extension of cohort analysis that has been suggested, e.g. by R. Jones in a paper presented to the 1973 ICES meeting, is to apply it to length–frequency data (see also Jones, 1981). In principle this avoids the need for large-scale age-determinations—which are not always easy—but does raise other problems, e.g. of changes in growth rate.

4.3.6 Estimation of fishing and natural mortality

The estimation of the total mortality tells us something about the dynamics of the population, but not very much about the effects of fishing until we know how much of this total is due to fishing and how much to natural causes. It is possible, as discussed later, to estimate directly the fishing mortality, and (with less reliability) the natural mortality. Alternatively—and this is probably the commonest method—the separation of fishing and natural mortality can be achieved by relating changes in the total mortality to changes in the amount of fishing.

Mathematically, we have

$$Z = F + M = qf + M \tag{4.30}$$

or if there are two periods with different, known, levels of fishing effort, f_1, f_2, during which estimates are made of the total mortality Z_1 and Z_2, we will have

$$Z_1 = qf_1 + M$$
$$Z_2 = qf_2 + M$$

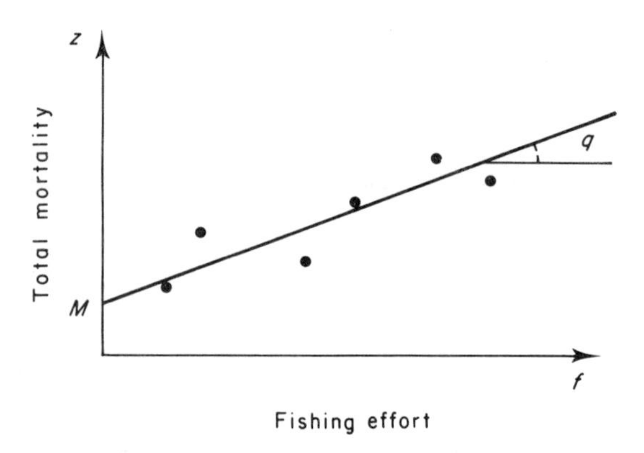

Figure 4.6 Separation of fishing and natural mortality by plotting total mortality, Z, against fishing effort f

This gives us a pair of simultaneous equations which can be solved for the two unknown quantities q and M. That is,

$$q = \frac{Z_1 - Z_2}{f_1 - f_2}$$

$$M = \frac{Z_1 f_2 - Z_2 f_1}{f_2 - f_1}$$

(see exercise 4.9).

More generally, we can see that equation (4.30) is in the general form $y = a + bx$, with the mortality Z being linearly related to the fishing effort f. Therefore, we can either proceed graphically, plotting Z against f (see Figure 4.6), and draw the best-fitting straight line, or calculate the regression line directly. (The wise man will probably do both, calculating the regression, and then plotting the result to see if there are points that fit the regression particularly badly, or if the regression is determined by just one or two unusual points.) In either case the slope of the line will give the catchability coefficient q, and the intercept on the y-axis the natural mortality M.

These estimates, particularly the latter, should be treated with a little caution. The true fishing effort is not, in many cases, accurately estimated by the available figures of nominal effort; this violates one of the basic assumptions of standard regression techniques. The practical effect is to flatten out the regression line, i.e. to result in an overestimate of M.

It should also be noted that only in certain situations (e.g. in the case of cohort analysis) is the total mortality during a year estimated. More often (see section 4.3.2), the mortality is estimated from the ratio of figures (e.g. c.p.u.e. of a given age-group) referring to average conditions during each of two successive years. In that case the mortality should be related to the average effort during the two years, rather than to effort in either of the two years, for example, in the graphical approach the total mortality between 1975 and

1976 should be plotted against the average of the effort in 1975 and 1976. This can be shown (e.g. Paloheimo, 1961) to make the most efficient use of the data.

In principle the method which is normally applied to sets of annual data of effort and mortality estimates, can also be applied to other situations in which fishing mortality varies, e.g. in a selective fishery. An example outside fisheries occurs in many game animals, in which hunting is often directed wholly or predominantly at males. Any excess in observed male mortality may therefore be ascribed to the effect of hunting. Most fisheries are selective to a greater or lesser extent usually towards fish of a given range of sizes. If the degree of selection is known, for example, that the fishing mortality on three-year-old fish is half that on six-year-olds, the difference in mortality at the two ages can be used to separate fishing and natural mortality. However, in practice the mortality rates will often be estimated from data of catches by the selective gear; this leads to a complex interaction between selection and apparent mortality.

Since the two quantities are not independent, estimation of natural mortality can lead to a circular argument, and to values that cannot be relied on. The application of the method requires independent estimates of age–composition in the population (and hence of total mortality and selectivity), for example from research surveys with a gear that can be supposed to be more or less non-selective over the range of ages in question.

4.3.7 Tagging

In the ideal world, tagging would be an excellent method for separating fishing and natural mortality rates, as well as estimating growth rates, migration, etc. In practice, difficulties in ensuring that fish are not seriously affected by the act of catching and tagging, that their behaviour after being tagged, as regards growth, migration, and mortality is typical of the population as a whole, and that any tagged fish that are later caught by fishermen are detected and reported to the tagging institution, have the result that only a minority of tagging experiments can be used with any confidence to estimate fishing mortality. The present manual, therefore, does not deal in great detail with the analysis of tagging data. The interested reader is referred to one or other of the more comprehensive reviews of tagging theory (e.g. Cormack, 1969; Seber, 1973; Jones, 1977). The practical problems of tagging are also reviewed by Jones (1979). A brief review will be given here of the general principles of estimating fishing mortality and a slightly longer discussion of the ways in which the assumption on which these principles are based may break down.

Tagging, or marking, is very widely used in studying populations of wild animals, from moths to whales. When the population being studied is small, the scientist can control the process of recapturing the tagged fish, for instance, by examining a sample of the population for the presence or absence of tags, or can arrange for tagging to be done in a succession of discrete batches, which may include, as a not uncommon event, the capture and marking of the same individual in more than one batch. Several of the mathematically most interesting and satisfactory methods of analysis are based on such multiple tagging. Scientists dealing with the more extensive populations which are subject of most commercial fisheries cannot hope to control their operations so closely. Once a fish has been tagged and released the scientist is most unlikely to see it again in his own catches. In practice, tagging experiments in fisheries which are the main target of this manual must be based on the return to the scientists of tagged fish (or information relating to these fish) which have been caught by the commercial fishermen.

In the simplest situation, a batch of 100 tagged fish may be released, of which say 60 are returned sooner or later. From this we might estimate that the exploitation rate (see section 4.3.1) was 60 per cent. Rather more information can be obtained by examining the time pattern of returns, for example the number returned in successive time intervals, of duration T, say. Then, noting that the population of tagged fish will tend to suffer rather greater reductions (losses of tags, deaths due to tagging, and migration out of the area, which cannot be matched by immigration of tagged fish), we can write the number of tagged fish present at the beginning of the rth period as

$$N_r = N_0 \exp[-(F+M+X)rT] \tag{4.31}$$

where N_0 = number tagged, and X = extra losses suffered by the tagged population. Hence the number caught during the rth time interval will be

$$n_r = \frac{FN_r}{F+M+X}(1 - \exp[-(F+M+X)T])$$

or

$$n_r = \frac{FN_0}{F+M+X}\exp[-(F+M+X)rT](1 - \exp[-(F+M+X)T] \tag{4.32}$$

or

$$\log n_r = -rZ'T + \log\left(\frac{FN_0}{Z'}\right) + \log(1 - e^{-Z'T}) \tag{4.33}$$

where $Z' = F + M + X$.

It can be seen from equation (4.33) that if the logarithm of the numbers returned in each interval is plotted against time, the result should be a straight line of the classical form $y = ax + b$, with a, the slope of the line, equal to $-Z'T$, and b, equal to $\log FN_0/Z' + \log(1 - e^{-Z'T})$. Once the regression line is fitted, then Z', and hence F can be readily calculated.

While these results can be obtained algebraically, it is desirable to make the actual plot. First the plot will reveal whether there is any departure from the straight line, which would suggest that the basic assumption of constant mortality is wrong. Second, since the variance will be roughly inversely proportional to the numbers returned in each interval, it will rapidly increase in the later intervals. The normal assumptions in fitting a regression line of constant variance is therefore violated, and short of applying a more complex procedure, a line fitted by eye, given greater weight to the earlier points, is likely to be acceptable. The graphical presentation also enables the important parameters to be read off quickly. The slope immediately gives Z', and if the values are plotted against the midpoint of the interval, and if $Z'T$ is not large, then the intercept on the y-axis, i.e. at the time of tagging, will be equal to FN_0T.

There are many ways in which the assumptions made in making the estimates can be broken, and the estimates of F and associated parameters of the population made invalid. These sources of error need to be carefully reviewed before the results of any tagging data are used in later stock assessments. To do so, it is useful to group them into those that occur during the tagging process, those occurring while the fish are in the sea, and those between the time a tagged fish is caught by a fisherman and the time when the information about that fish reaches the scientist analysing the tagging data (if in fact he ever gets it).

The assumption often made about the release operation is that if 1000 fish are caught, brought on deck, manhandled by a scientist who may be seasick and clumsy, have a tag stuck into them, and dropped over the side, then a few hours later there will be 1000 healthy fish swimming around, normal in every way except that they are each carrying a tag. This may be so if the fish are very hardy, or the methods of catching and tagging gentle, but usually it must be accepted that the number of fish effectively tagged (and the number N_0 that should be used in the above equations) will be less than the number of tags applied.

It is not intended here to discuss the practical methods of catching and handling fish, and the choice of tags which will be most likely to reduce the losses at tagging (see, for example, Jones, 1979). The concern here will be methods of analysis to detect, and as far as possible, make corrections for, losses at tagging. This may be approached by dividing the fish into different groups that might be expected to be differentially affected by the shock of capture and tagging. This may be done by subjective judgement, for instance by classifying the fish as lively or sluggish, or according to more objective criteria, e.g. the duration between capture and release, weather conditions, or even according to the scientist or technician who actually tagged the fish (see exercises 4.13 and 4.14).

If there is some mortality at tagging, but some fish do survive, then it is reasonable to suppose that the percentage surviving will vary according to the precise tagging conditions. It follows that a comparison of the percentage returned of fish of different conditions will give some insight into the possible losses at the time of tagging. It should be stressed that the definite information is in one direction only. That is, if (after due allowance is made for statistical probabilities) a smaller percentage of fished tagged under poor conditions are returned than of fish tagged under good conditions, this shows that some of the former group did not survive the tagging process, and provides an upper limit to the proportion that did survive. The comparisons, however, cannot show what proportion, if any, of the most favoured group failed to survive.

If (again after due allowance for chance differences) there is no difference between groups, this can be taken as suggestive of low tagging losses, but it is not a definite proof.

Mathematically, if N fish are tagged, in k different categories, N_r being classed in the rth category, $\Sigma N_r = N$, the first category being believed *a priori* to be the group most likely to survive, and n_r are returned from the rth category, then we can calculate and compare the proportions $p_r = n_r/N_r$. If these are statistically different, then taking the first group as standard, the number in the rth classes that would have had to be tagged successfully to give the observed number of returns is $N'_r = n_r/p_1$; thus the estimated proportion of losses is $1 - N'_r/N_r = 1 - p_r/p_1$.

It is easy to introduce bias into this analysis, especially when dealing with small numbers, by selecting, after the experiment, the class which, by chance, happens to have had the highest proportion of returns. The correction for possible losses should be based on the class expected, *a priori*, to suffer least mortality (unless there is statistically significant evidence that these expectations were wrong), and other classes can be included if they have as high a proportion of returns.

Using the corrected number of effective releases, a total number $N' = \Sigma N'_r$ can be calculated, and this should be used as the effective number tagged when applying, for example, equation (4.33).

A similar approach is possible to make partial corrections for incomplete returns. Different groups of fishermen may vary in their interest in and awareness of the tagging work and in the way they handle the fish, so that the probabilities of their detecting a tag, and if detecting it, returning it to the scientists, may differ. These probabilities may be measured (with due allowances for differences in areas fished, size of fish caught, etc. relative to the area of tagging, etc.), by the number of tags returned per unit weight of catch. If these differ—especially if the direction of the differences seem reasonable in the light of what is known about the fishermen and their interest in tagging—corrections can be made to obtain estimates of the number of the tagged fish that were actually recaptured, on the assumption that the best group of fishermen detected and returned all the tagged fish caught (see exercise 4.15).

The proportion of returns may also be estimated directly by introducing a known number of tagged fish into the catches. This is rarely practicable for normal external tags on fish used for human consumption without altering the awareness of the fishermen, but can easily be done for fish used for fish meal. These are usually tagged with internal tags, which are later recovered on the magnets in the factory. Testing the effectiveness of the magnets at each factory should be a routine part of any experiment in a fish-meal fishery.

Differences between the group of tagged fish in the sea and the population as a whole often present major difficulties. One kind of difference has already been touched in equation (4.31), where a loss rate X has been introduced in addition to the usual fishing and mortality rates. This loss rate includes possible increased mortality from predators or disease due to carrying a tag (the scientist is often faced with a choice between a highly visible tag which at least on small fish may well increase predation, and a less visible tag, which may be missed both by predators and by fishermen, so that few are returned), as well as the loss of the tag from the fish. The latter may be detected, and a loss rate estimated by attaching two tags (preferably of different types) to the same fish, and noting when only one tag is returned. For example, following Gulland (1963) (and correcting a misprint in his equation), if the probabilities of one or other of two tags becoming detached after time t are ${}_{A}p_t$, and ${}_{B}p_t$, the numbers of fish bearing only one tag of one or other type, expressed as a proportion of fish still with both types can be written as

$$\frac{{}_{1}N_{A}}{{}_{2}N}=\frac{{}_{B}p_t}{1-{}_{B}p_t};\qquad \frac{{}_{1}N_{B}}{{}_{2}N}=\frac{{}_{A}p_t}{1-{}_{A}p_t}$$

From these expressions which, while referring to fish in the sea, can be estimated from the proportion of the different types in the returns, the proportion of tags that have become detached can be readily estimated. Further, by plotting the proportion of fish with single tags as a function of time it can be seen whether or not the loss of tags can be considered as a constant exponential rate, similar to most other causes of mortality (as is implied by the constant coefficient X). In practice it is likely that the additional losses in the tagged population will be initially high (e.g. due to delayed shock of tagging, etc.), then moderately low, and then increase as the fish grow or the tags wear out; nevertheless a constant coefficient may be an adequate approximation.

The other big difference between the tagged and untagged population lies in their geographical distribution. For practical reasons tagging has to be concentrated at a few points, usually where fish can be caught in good quantities; it is virtually impossible to spread the tags randomly through the fish population. Since commercial fishing is also non-randomly distributed, and concentrated where catches are best, the fishing mortality on the tagged fish will be different from, and usually higher than, that on the population as a whole. If mixing is relatively quick, returns some initial period, 0 to t', during which the tagged fish are mixing with the untagged population, may be omitted in the computations. The analysis is then carried on from time t', with a reduced initial number of tags, using equation (4.31) in the form

$$N_t = N_0' \exp[-(F+X)(t-t')] \tag{4.34}$$

N_0' is estimated from the number tagged, subtracting the number returned in the initial period of mixing, and the estimated number lost through other causes, X. This last estimate will have to come from an analysis of the later data. However, in many situations nearly all returns will be made before the tagged and untagged populations are even approximately mixed.

This mixing may be effectively speeded up by a suitable pattern of tagging. The aim is to have the ratio of tagged to untagged fish the same throughout the population. This might, for instance, be achieved by making an evenly spaced grid of trawl stations covering the whole area, and tagging all, or a constant proportion of, the fish caught at each station. This is possible in a small area but not easy to do in, say, the North Sea.

In a large area such as the North Sea it may not be possible to ensure effective mixing of tagged and untagged fish, or at least by the time the mixing has occurred most of the returns have taken place, and to ignore returns occurring before mixing was effectively complete would mean discarding most of the available data. However, if good information is available on the geographical distribution of the fishing effort, and on the positions of the capture of the returned tages, useful estimates of the fishing mortality on the untagged population can be made. The logical steps are:

(a) Estimate the fishing mortality, F, on the tagged population, e.g. from equation (4.31).
(b) From the distribution of fishing effort and of the tagged fish estimate the fishing intensity, f, on the tagged population (see section 2.3.4).
(c) Hence, estimate q, in the relation $F = qf$.
(d) Estimate the fishing intensity on the whole population.
(e) Hence, assuming that q for tagged and untagged fish is the same, estimate the fishing mortality on the whole population.

It should be noted that, especially in the first steps, it is not necessary to consider fishing by all vessels, or in all areas, provided the same section of the total is used in computing fishing mortality (from returns of tags) and fishing intensity (from data on fishing effort). For instance, in the North Sea some of the most extensive detailed data on fishing effort is of fishing by United Kingdom deep-sea trawlers; therefore, it may be convenient to compute the fishing mortality and intensity only from the data of United Kingdom trawlers, and ignore returns of tags, and fishing activity by United Kingdom seiners and inshore vessels, and by vessels of other countries.

The procedures of (a) and (c) above can be combined to give a direct estimate of q, as follows.

Consider a small area around the position of tagging in which mixing of tagged and untagged fish takes place quickly. For a certain time interval, i, let

$\bar{N}_i$ = mean number of tagged fish present

f_i = fishing intensity (effort per unit area per unit time) as determined from detailed data on fishing effort and its distribution

n_i = number of tagged fish returned

Then the returns per unit effort will be proportional to the numbers present, i.e.

$$\frac{n_i}{f_i} = q\bar{N}_i \tag{4.35}$$

If then n_i/f_i, the tags returned per unit fishing intensity, is plotted against time and a curve fitted to the points, the intercept of this curve on the y-axis will be qN_0, where N_0 is the number released. Hence, q can be estimated and the fishing mortality on the population as a whole will be given by

$$F = q\bar{f}$$

where $\bar{f}$ is the effective overall fishing intensity on the populations, as calculated from detailed effort statistics.

4.3.8 Other estimates of fishing or natural mortality

In addition to the methods described in the previous sections, which, if the basic data are good enough, should provide quantitative estimates of fishing and natural mortality, there are a number of other approaches which are less demanding on data, though are less likely to produce precise estimates. Since the fishery scientist is seldom lucky enough to have as good data as he would like, it is always worthwhile looking at the alternative approaches discussed here. Even if the basic data seem good, estimates of mortality from independent sources may, if different, suggest unsuspected sources of error in, for example, tagging results or, if similar to those obtained earlier, give welcome support to those results.

Direct census

If the abundance of the total stock is known, either at some particular point in time, or as an average over a period, and the catch is known, the exploitation rate and fishing mortality can be calculated, e.g. from equation (4.25). In a few cases direct counts can be made visually, for instance of salmon passing upstream (electronic methods can also be used in this case, EIFAC, 1975) or of gray whales passing during migration off southern California. Visual methods of counting can also be used, on a sample survey basis for marine mammals over a wider area (e.g. Doi, 1974). Modern acoustic methods enable quantitative estimates of abundance to be made for a wider range of species (Forbes and Nakken, 1972; Burczynski, 1979), but conditions have to be favourable—little difficulty in species identification, and with the fish distributed neither too close to the bottom nor too close to the surface—for the quantitative estimate to be entirely reliable and valid for use in estimating fishing mortality.

In many ways the most satisfactory organisms to count in the sea are fish eggs (and to a less extent young fish larvae). They cannot dodge the nets, nor are they so small as to pass easily through the meshes of a normal plankton net. Therefore, plankton surveys have been used to estimate the number of eggs or larvae (Smith and Richardson, 1977), hence the numbers of adult females, and hence (from the catches of adult females) the fishing mortality on them, which may then be assumed to be the same as that on the exploited part of the population as whole. Two major difficulties in this approach are the identification of eggs and, to a lesser extent, early larvae, and the patchiness of spawning and hence the high variance associated with a reasonable intensity of sampling (e.g. English, 1964) (or, put another way, the large number of observations and hence high costs necessary to achieve a reasonably small variance in the estimates of total egg numbers, or number of adult females). The result is that the fishery scientist will seldom find it advisable to allocate a major part of his limited budget to a survey of fish eggs or larvae specially directed to estimating the spawning biomass (though such surveys may be carried out as part of a research project with wider interests), but should take every opportunity to take advantage of egg or larval data collected as part of other programmes.

The total biomass may also be estimated from surveys by ordinary fishing gears. The commonest gear used is bottom trawl, and the usual assumptions are that the area sampled by the survey is a random sample of the total area inhabited by the stock (or within appropriate stratifications), and that the trawl catches all, or some specified proportion, of the fish in its path. These assumptions, particularly the latter, are probably not too well satisfied, so that only approximate values of biomass, and hence of fishing mortality, can be obtained in this way. Nevertheless, since the assumptions are independent of those made in using other approaches, these estimates can be useful in producing an additional indication of where the correct value lies.

Swept area

The same assumptions (random distribution, and catching all fish in the path of the net) are involved in another approach, using the area covered by the activities of the commercial fleet. If the assumptions hold, and during unit time the gear used by a given vessel covers an area, a, out of a total area A inhabited by the stock, then the fishing mortality caused by that vessel is a/A. Adding together the activities of all the fleet, covering an area a', the fishing mortality will be a'/A. The precision of this method is low; fishermen will always concentrate on the higher densities, and for many gears (e.g. traps, gill-nets) the area 'swept' by the gear is poorly defined. The main value of the method is that it makes little demand on data, either from commercial statistics, or research work, and therefore can give a quick first approximation at an early stage of the study of a fishery. It can also be used, and should indeed be used quite widely, to give a common-sense check to results obtained by more sophisticated methods. If the stock covers a wide area, is rarely concentrated, and only a few boats operate, then the fishing mortality can hardly be high, but if the stock inhabits a small area, or is regularly concentrated (e.g. for spawning), and vessels are crowded on these grounds, then the fishing mortality is almost bound to be high.

Natural mortality and growth

Both the natural mortality rate of a fish and the curvature in its growth curve (i.e. the rate at which it approaches its maximum size) are measures of the rate at which the fish lives. They can therefore be expected to be related; a long-lived fish will approach its limiting size slowly, whereas a short-lived fish will, in this sense, grow fast (though since the long-lived fish are often big, their absolute growth rate, expressed as centimetres per year, may be higher). Since, in the von Bertalanffy formulation of the growth pattern the parameter K describes the rate at which the maximum size is approached, we might expect M and K to be closely related. Examination of empirical data shows that in fact M and K tend to be proportional, with a constant of proportionality that is different for different taxonomic groups of fish. For clupeoids M is generally between one and two times K, whereas for gadoids M is between two and three times K (e.g. Beverton and Holt, 1959). This apparent constancy (within broad limits) of the ratio M/K for different groups is particularly helpful in the calculation of yield curves where, as is discussed in section 5.2, the ratio M/K tends to appear as a single explicit parameter, rather than M and K separately.

4.4 RECRUITMENT AND SELECTION

4.4.1 General considerations

Section 4.3 considered the problems of estimating the magnitude of fishing and natural mortality, without much concern about how the mortalities might differ between fish of different ages, except as an unwelcome complicating factor in the estimation procedure. In practice, even apart from the obvious fact that normal fishing gear cannot catch or retain very small fish, fishing mortality can be very different on fish of different ages (or sizes), and changes in the age-specific fishing mortality (especially reducing it on the smaller fish) often provides opportunities for managing the fishery which can be put into effect more easily than controls on the magnitude of the fishing mortality (or fishing effort). Where these opportunities may exist the estimation of the pattern of changes in fishing mortality with age, and of the effects on catches of alterations in this pattern (see section 5.3) deserve early attention.

It is often convenient in discussing how fishing mortality varies with age to distinguish between recruitment and selection. The former is a characteristic of the fish themselves, and is the process whereby the young fish, previously inaccessible to the ordinary fishing gear become as a result of growth, change of behaviour or of movement on to the fishing grounds, potentially vulnerable to fishing; the latter is a characteristic of the fishing gear itself and the way it is operated.

Conceptually, the process of recruitment, and the separation of the entire population of a species into two groups—the pre-recruits (including eggs, larvae, and, usually, juveniles), and the post-recruits, or exploited phase of the population—is an important and usually a useful one. It is only the latter group of fish that can easily be studied by the methods of this manual. Indeed, to fishery science in the narrowest sense the pre-recruit phase can be considered as a black box, with a certain number of adult fish providing the input and the output appearing as a certain number of recruits. It has become clear that what happens in this black box, especially how the output is related to the input, is extremely important to the understanding and management of fisheries (see Chapter 6). While it will remain convenient for many purposes for the fishery scientist to continue to talk about the exploited phase as though it were the entire population—and he may continue to refer, for example, to cod as being fish between 30 and 120 cm in length, feeding mostly upon smaller fish—the fact that, in terms of numbers at least, most cod are small, often planktonic, animals feeding mostly on zooplankton and liable to be preyed upon by, for example, mackerel, may have to be recognized to an increasing extent, especially when considering possible interactions between different species (see Chapter 7).

The process of recruitment can involve a clearly defined and observable event, such as substantial movement or migration from 'nursery areas' (e.g. of penaeid shrimp from coastal lagoons to offshore trawling grounds), or change of habit (e.g. the settlement on the bottom of young cod or haddock which for their first few months live in mid-water), but may involve no more than growing big enough to become of interest to the fishermen. It may be noted that the movement of shrimp, by which they recruit to the offshore fishery (typically carried out by moderately large commercial trawlers) may also remove them from the artisanal fisheries in the lagoons, i.e. what is the process of recruitment from one point of view is also from another point of view emigration of animals from a fishery to which they recruited some months earlier. Thus recruitment, though basically a biological process, is not independent of the type of fishery.

Selectivity, in a similar fashion, while basically a characteristic of the fishing gear, is also, at least when considered in a wider sense, dependent on the behaviour of the fish. In the narrow sense, selection consists of the differential escape of certain sizes of fish after they come into contact with the gear (for example the passage of small fish through the meshes of a trawl). However, as usual, matters are not all that simple. Where good data exist of the variation of fishing mortality with age, e.g. from cohort analysis, it is clear that fishing mortality may be far from being constant, even if there are no obvious physical characteristics of the gear that might cause selection in the narrow sense. This is perhaps not surprising. On the one hand, fishermen do not fish at random, but make every effort to fish where and when catches are most rewarding; on the other, fish are not distributed at random, and the distribution of different sizes or ages may be different. Obvious examples are spawning concentrations (so that the fishing mortality may, even when taken over the year, be much higher on mature fish) and shoaling fish, which often school by size, so that purse-seines and other gears, which may appear non-selective, can in fact exert a highly selective mortality on certain sizes . This effect probably occurs in most fisheries, even those in which there is no obvious cause for differential mortality. For example, Table 30 of Bannister (1977) shows that the fishing mortality on 4–6-year-old male plaice in the North Sea may be twice that on 10–15-year-old males, or on 4–6-year-old females, even though all these fish would appear to be fully recruited and well beyond the selection range of the gear used.

Consideration of the possible patterns of variation with fishing mortality with age arising from all causes–and not just the recruitment of young fish and the escapement through meshes and other simple forms of selection—and as far as possible its estimation, should receive careful attention. Unfortunately, this is not always easy. Section 4.3 has shown how difficult it can be to obtain any estimate of fishing mortality, let alone its variation with age. Where the available data allow, cohort analysis provides direct estimates of the fishing mortality on each age-group. Otherwise tagging provides an approach that can, in principle, give estimates of the variation of mortality with age.

If fish of different sizes are tagged, any difference in the rate of return of tags should provide an estimate of difference in fishing mortality. In practice the difficulties of distributing tags throughout a large population so that any differences are characteristic of the whole population, rather than that part of it in the tagging area, plus the possibility that any differences actually observed may reflect the differences in any additional mortality due to capture and tagging, makes this approach of less than general applicability.

The most widely applicable approach, if only qualitative, is to examine the size or age composition of catches by different sectors of the fishery (different gears, vessels landing in different ports, etc.). If these are different, then at least one sector must be selective. Examination of the nature of the gears involved, the behaviour of the fish, and the strategy and tactics of the fishermen, may suggest which sector (or sectors) are most likely to be non-selective. The fishing mortality on the *i*th age (or size) group will then be proportional to C_i/C'_i, where C_i is the total catch by the whole fishery of that size group, and C'_i is the catch by the non-selective gear.

Fishing mortality can vary, due to recruitment or selection, over a range of ages (or sizes), but for purposes of discussion and analysis it is easier, and often acceptable in practice, to treat recruitment or selection as being abrupt or 'knife-edged'. That is that no fish enter the fishery until they reach an age t_r, but all fish of age t_r and older are fully recruited. Similarly, selection may be supposed to act so that fishing mortality is zero until the fish reach the mean selection age t_c, and therefore undergo the full, constant fishing mortality. As discussed below this seems to be an adequate simplification for trawl selection, but for other types of gear, e.g. gill-nets, a more complex model is needed.

4.4.2 Trawl selection

Selection is simplest in the 'bag' type of gear—trawls, seines, etc. For these gears it is usual to assume that the size composition of the fish entering the mouth of the net is the same as that in the immediate vicinity of the gear. The selectivity of such gears therefore becomes a question of escape through the meshes of fish which have entered the net. For many species there is evidence to show that most of this escape occurs through the cod-end. Selectivity can therefore be determined directly if the numbers of each size of fish entering the net can be estimated, either by attaching a small-meshed cover over the

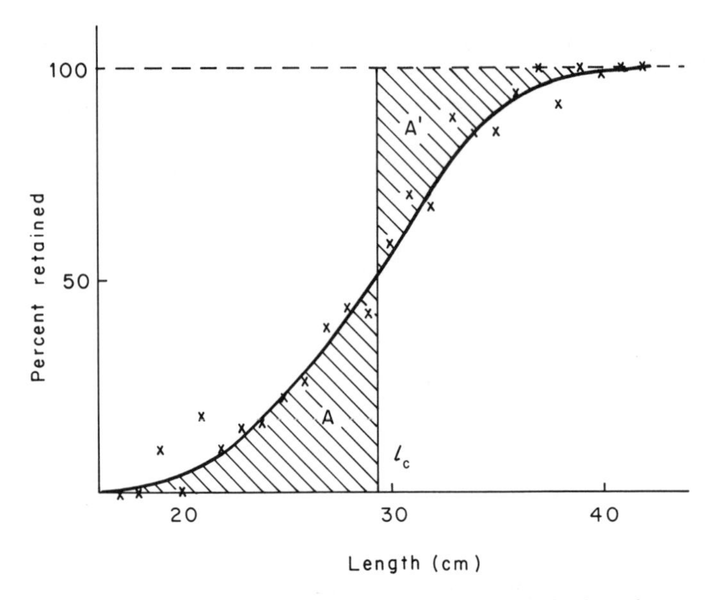

Figure 4.7 Selection of hake by a 77 mm cod-end (data from Gulland, 1956, reproduced by permission). The curve has been drawn by eye. The mean selection length l_c is chosen to equalize the shaded areas A, A′

cod-end or other parts, or from the size – composition of the catches of nets of much smaller meshes fished at the same time and place. Whichever method is used, the results can be expressed as the proportion of fish at each length entering the net which are retained in the cod-end. When these proportions are plotted against length, the selection curve of the net for the species concerned is obtained. A typical example is shown in Figure 4.7.

A selection curve may extend over a range of length of fish of perhaps 10 cm or more, which means that as young fish begin to grow into the selection range they at first suffer only a low fishing mortality. As they grow larger, their chance of escaping from the net (having entered it) gets less, until, eventually, they are too large to escape at all; only then are they exposed to the full fishing mortality rate—whatever that may be. Although it is possible to introduce this progressive change of fishing mortality coefficient over the selection range into yield assessments, it is usually sufficient to represent the selection process by a single mean selection length l_c, and assume that all fish of length less than l_c are released, and that all those greater than l_c are retained by the gear, and experience the full fishing mortality.

The important quantity is therefore

$$t_c = \text{mean selection age}$$
$$= \text{mean age of entry to the catch, or mean age at first capture}$$

In Figure 4.7 the mean selection length l_c is the length which makes the two shaded areas A and A′ equal. If the selection curve is symmetrical, or nearly so, this will be the length at the midpoint of the curve, i.e. the 50 per cent length at which half the fish entering the net escape and half are retained. If the curve is not symmetrical, l_c can be calculated by equating the two areas between the selection curve the y-axis, and the line $l = l_c$. Suppose the lengths are classified, the ith class being of size h_i and limited by the lengths l_i and l_{i+1}; the corresponding ordinate to the ith class is y, where $i = 0, 1, 2, \ldots n$.

Then the area to the left of the selection curve is equal to the area of the rectangle of height 1 and base l_{n+1} minus the area under the selection curve. This last area can be approximated by the sum of the rectangle areas for each length-class, i.e. $\Sigma h_i \cdot y_i$. The total area to the left of the selection curve is therefore $l_{n+1} - \Sigma h_i \cdot y_i$ which, as the total height of the curve is 1, will equal l_c. If h_i is constant and equal to 1 cm, $l_c = l_{n+1} - \Sigma y_i$.

For trawls, the mean selection length l_c is generally proportional to the mesh size, m, i.e.

$$l_c = b \cdot m$$

where b is the selection factor. This expression enables the mean selection factor for any mesh size to be estimated from the results of experiments with a small number of different mesh sizes. This may best be done by plotting mean selection length against mesh size, and drawing the best-fitting proportional line through the points. There is some variation in selection factor with the condition of the fish, and also with towing speed, and with the material and construction of the meshes; light flexible materials (e.g. nylon) give higher selection factors than thicker materials (e.g. manila). There may also be variations with the size and nature of the catch. Fish may find it harder to escape from a large catch, or a catch with a lot of rubbish—weed, etc.—thus giving a lower selection factor. It is therefore important to carry out mesh selection experiments under conditions that correspond as closely as possible to those in the normal commercial fishery. It is often convenient to carry them out on board a chartered commercial trawler. Since the selection depends on only the physical dimensions of the fish and the mesh, selection will be the same for a given species in any part of the range; this means that (unlike estimates of growth and mortality) someone working on one stock does not need to repeat selection studies carried out on a different stock of the same species. Indeed, it can be expected that fish of similar shapes will have similar selection factors, and thus approximate selection factors can be obtained for several species once that for one species of the same shape has been estimated. This can be a useful way of obtaining quick answers in tropical areas where there are a large number of species (see Jones, 1976).

An aspect of selection studies, and even more of the enforcement of regulations on mesh size, that is sometimes neglected is the actual measurement of the size of the mesh. While there are various ways of defining the size of a mesh (bar, length, number of rows per yard), the standard measure, at least in respect of trawl nets, is the internal stretched diameter. This will be half the perimeter of the hole through which any released fish will have to pass. The practical difficulty in measuring this is that some force is needed to pull the mesh out flat, but this force will also tend to stretch the twine or distort the knots, so that a larger measurements is obtained. The mesh should therefore be measured with some type of gauge that provides a standard force, (see Pope *et al*, 1975, section 3.8).

4.4.3 Gill-net selection

Gill-nets are about the most selective gear used in commercial fishing. If a gill-net is used on a population with a fair spread of sizes the length–composition of the catch will be largely determined by the selectivity of the net, and will have a shape like that of curve A in Figure 4.8. While differences in the pattern of mortality caused by two trawl nets differing only in mesh size can be described, in theory by a single quantity (the mean selection length, l_c, though in practice trawls with larger meshes may, due to better flow through the net, catch more fish above the selection rage), two gill-nets with the same overall dimensions can differ in the pattern of fishing mortality caused in at least three ways—the spread of the selection curve, the position of the peak of the curve, and the height of the peak.

These are illustrated in Figure 4.8. The fall-off in catches each side of the peak are because the net has its peak effectiveness only for one size, failing to retain the bigger fish and allowing the smaller ones to pass through. A net with similar mesh sizes, but made of somewhat different and more flexible material, may not lose its effectiveness quite so rapidly, and thus retain more fish each side of the

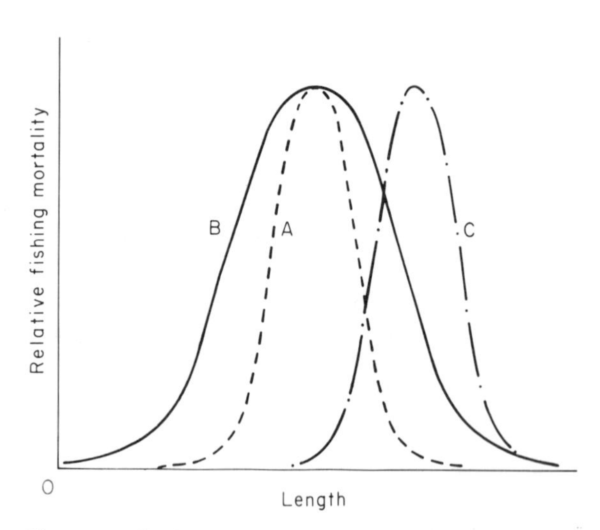

Figure 4.8 Typical selection curve of a gill-net (A), showing also the selection curve of a net (e.g. with more flexible material) with a broader range (B), and of a net with the same range, but a bigger peak selection length (e.g. with a bigger mesh size (C))

optimal size, as shown by curve B. A similar net, but with larger meshes, may have the same curve, but shifted up to a larger optimum size (curve C). Usually, as in the case of trawl selection, it can be assumed that the optimum length l_m is proportional to the mesh size m, i.e. following the terminology for trawls in section 4.4.2 we can write $l_m = bm$, where b is the selection factor. (It may be noted, e.g. Hamley, 1975, that in the gill-net literature a selectivity factor K, defined by $m = Kl_m$ is commonly used; since mesh size is usually the independent variable, the present terminology seems preferable.)

In practice a larger mesh may not exert the same mortality on the optimal size of fish as does a net with smaller mesh. Since capture must be preceded by the active movement of the fish to encounter the net, and larger fish might be expected to be more active and to move over longer distances, their probability of encounter, and hence of retention if they are of the optimal length for the mesh concerned, will be larger. This expectation tends to be confirmed by practical observations; therefore the peak in the selection curve of a larger mesh will be higher.

These effects make estimation of gill-net selection a more difficult matter than estimating trawl selection. Fortunately, gill-nets are often used in small bodies of water in which direct methods of estimating mortality can be more easily used than in the open ocean. For example, it may be comparatively easy to spread tagged fish randomly through the population and compare the recapture rate of different sizes of fish (taking care that the type of tag used does not alter the probability of capture). It may even be possible to estimate the absolute number of each size of fish in the body of water concerned. Failing this it may often be possible to compare gill-net catches with those taken by less selective gears, such as trawls or purse-seines, and thus, with due reservations about the selectivity of the other gear, estimate the selectivity of the gill-net.

However, often the selectivity will have to be estimated from catches of gill-nets with different meshes. In doing this the first step is to note that the entire process can be described by a three-dimensional diagram (e.g. Figure 17 of Hamley, 1975), giving the fishing mortality caused by unit effort (e.g. standard area of net) as a function of the mesh size and the length of the fish. This diagram can be looked at from two directions; with fixed mesh size, to give the relative mortalities on different sizes, and for a given fish size the relative mortalities caused by different meshes. The former (Type A curves of Hamley) is the aspect of selection that is usually considered, and is the important aspect for many purposes, for example when interpreting the results of catches taken by a single mesh size (or some combination of mesh size) to provide some estimates of the true composition of the population of fish in the sea (or lake or river). The latter aspect (Type B curve of Hamley), however, is important in some situations, for example when considering the effect of a change in mesh size (or other changes in the nature of the gear, e.g. in the type of twine used).

This is fortunate because this aspect can be readily observed. All that is needed is to fish nets of different sizes together and compare the catches taken by each net, doing this separately for each size group that occurs in significant quantities.

The plots (of numbers caught against mesh size) for each size group can be treated separately, but there are obvious advantages in being able to look at the curves as a whole, by expressing them in some standard form. So far as the x-axis is concerned this standardization can be achieved by looking at the relative dimensions of mesh and fish. Theoretically this might be done by calculating, for each mesh size and size group, the ratio of maximum fish girth to mesh perimeter (e.g. McCombie and Fry, 1960). However, in practice girths are seldom measured directly, and are usually deduced from the length. It may therefore give a fairer idea of the basic data to use the ratio of fish length to mesh perimeter, or nominal mesh size.

Standardization of the y-axis is more difficult. The numbers caught will be the product of the frequency of the length group in the population and the fishing mortality caused by the particular mesh size, of which the former at least will differ greatly between length groups. One approach, following McCombie and Fry (1960), is after fitting the curves for each length class to adjust the scale of the y-axis to equalize the area under each curve. Alternatively the semi-graphical method of Ishida (1962) can be used; by plotting the logarithm of the numbers caught, the proportional factor due to differences in the population numbers in each length class appears as a constant vertical difference in the graphs. If these are plotted on transparent graph paper the vertical differences can be adjusted to make the curves coincide. The amount of vertical movement required to achieve this gives a factor by which the observed numbers should be divided by to bring the Type B curves for each length class on to a common scale.

These standardization procedures, expressing the mesh size as some function of the length or girth of the length class of fish being considered, and then adjusting numbers caught by that mesh by some factor common to all catches of that size class, enable all the data to be plotted in a single Type B curve. This has two advantages apart from the obvious one of illustrating all the data in a single compact form: first, it enables a test to be made of the implicit assumption that the curves for all size classes are in fact similar, and can be expressed after transformation in the same form; second, provided the assumption appears consistent with the data, the combination of all the data make it much easier to determine accurately the form of the curve. While points corresponding to each mesh size on the Type B curve for any particular size group can be determined directly, they will be few in number—it is a rare experiment that uses more than half a dozen mesh sizes, and several of these may be too big or too small to obtain significant catches of any given size group—and often insufficient to determine the form of the curve without other input.

Determination of the Type B curves for each size group does not in itself determine the Type A curve for any mesh size. This is true even though when transformed (e.g. to show relative catches as a function of the girth: perimeter ratio) the two types of curves may look the same. It is not necessarily true, for example, that because for a given size of fish, a mesh size 5 per cent smaller than the optimum will exert a fishing mortality half that exerted by the optimum mesh, then, a given mesh size will, on a fish 5 per cent bigger than the optimum, exert a mortality half that on the optimum fish size. This is a reasonable assumption, and if it is made then Type A curves can be obtained directly from Type B curves. Other assumptions are possible, for instance, that the mortalities caused by any mesh size on the optimum size of fish for that mesh size are the same, which will enable Type A curves to be deduced.

A particularly useful set of assumptions are those used by Holt (1963) to determine the selection curves for a pair of meshes. The basic assumption is that the selection curve is normal, e.g. that the catches by a net of mesh size, m, of fish of length l may be given by

$$_{m}C_{l} = F_{m}N_{l}\exp\left(\frac{-(l-l_{m})^{2}}{2S_{m}^{2}}\right) \tag{4.36}$$

where $l_m =$ length for which the mesh size is most effective, $S_m^2 =$ variance of distribution $N_l =$ numbers of fish length l in the population, and $F_m =$ fishing mortality on fish of length l_m.

To compare the catches of two mesh sizes m_1, m_2 will assume further that l_m is proportional to the mesh sizes, i.e. $l_1 = bm_1, l_2 = bm_2$, that $S_1 = S_2 = S$, and, that if the nominal effort (e.g. number of standard-sized nets used) with each mesh size is the same, $F_1 = F_2$.

Taking logarithms we have, from the equation above,

$$\begin{aligned} \log(_{1}C_{l}) &= \log FN_{1} - \frac{(l-l_{1})^{2}}{2S^{2}} \\ \log(_{2}C_{l}) &= \log FN_{1} - \frac{(l-l_{2})^{2}}{2S^{2}} \end{aligned} \tag{4.37}$$

or taking the ratio, and substituting for l_1, l_2

$$\log\left(\frac{_{1}C_{l}}{_{2}C_{l}}\right) = -\frac{(l-bm_{1})^{2}}{2S^{2}} + \frac{(l-bm_{2})^{2}}{2S^{2}} = \frac{b(m_{1}-m_{2})}{S^{2}}l - \frac{b(m_{1}^{2}-m_{2}^{2})}{2S^{2}}$$

That is, if the logarithm of the ratio of the catches are plotted against length, the result should be a straight line, slope $b(m_1^2 - m_2^2)/S^2$ and intercept $b(m_1^2 - m_2^2)/2S^2$. After making the plot (to see if the points indeed fall on a straight lines), the slope and intercept can be determined at once graphically or by normal statistical computation. From these the two unknown quantities b and S can be estimated, and thus the selection curves determined.

When the selection curve of the gill-net has been determined, subsequent population analysis is simpler if the curve can be represented by a more simple form, analogous to the knife-edge approximation to the trawl selection. Two selection lengths are needed, l_c, the length at which the fish enter the selection range, and l_d, at which they grow out of it. That is, the fishing with a gill-net with the selection curve of Figure 4.9 has the same effect as a constant mortality

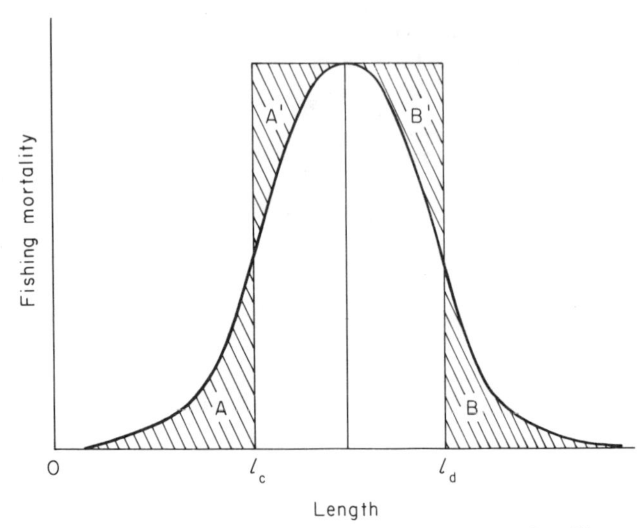

Figure 4.9 Approximation to the selection curve of a gill-net. The shaded areas A, A′, and B, B′ are equal

between lengths l_c and l_d, where the pairs of shaded areas A and A′ and B and B′ are equal.

4.4.4 Selection by other gears

The previous sections have discussed the selectivity of trawls and gill-nets at length. This is because the scientific literature on selectivity has been mainly concerned with these gears, because most of the practical controls on the use of selective gears have dealt with the legal requirements on the sizes of trawl or gill-net meshes, and because experiments to measure the selectivity of these gears can be readily designed and executed. All other types of gear are also, at least to some extent, selective and the pattern of selection can often be changed by suitable adjustments to the gear.

In many cases selectivity will be governed by the physical characteristics of the gear, for example, a given size of hook will only capture efficiently fish that are big enough to get the hook into their mouth, yet not so big that the hook cannot hold it. Estimation of the selectivity of these gears can in general follow the methods described in the previous two sections. The first step is to fish with gears in which the characteristics, e.g. hook sizes are allowed to vary. If, as in the case of hooks, the selectivity is likely to be optimum for a particular length, and to fall off above and below that size, the methods described for gill-nets can be applied, with obvious modification of detail. If the selection pattern is to allow fish below a certain size to escape and larger ones are retained, the trawl model should be followed. An example of this is the escape gaps used in the Western Australian rock lobster fishery (Bowen, 1963; 1971).

Selection can depend on more than just the geometry of the gear, and may vary with the characteristics of the population. For example, it has been suggested that the presence of large lobsters may discourage smaller lobsters from entering a pot. While there are many large lobsters in the population, the effective fishing mortality on small ones may be low. Even trawling can have such complications in that the effect of the bridles in shepherding fish into the path of the trawl may be greater for large fish than for small. Some of these complications cannot be dealt with by direct experiment. Other approaches, such as looking at the fishing mortality rates at different ages, as estimated from cohort analyses (see section 4.3.4) may be used.

4.4.5 Combination of selection with other processes

If selection and recruitment are occurring over the same ranges, as in Figure 4.10, then the effective selection, i.e. the proportion P_1 of the full fishing mortality to which the fish of a given size in the stock are exposed, will be given by the equation

$$P_1 = r_1 \times S_1 \tag{4.38}$$

where $r_1 =$ proportion recruited, i.e. ordinate of recruitment curve, and $S_1 =$ proportion of those entering the net which are retained, i.e. the ordinate of the selection curve. Thus the resultant curve, which expresses the effective entry of fish to the catch, is obtained as the product of the recruitment and mesh selection curves, and from it the resultant mean selection length can be determined as before by equating the two areas A, A′.

Sometimes, as in cases where the recruits migrate into a fishing area, recruitment, in terms of length, can be quite sharp, approaching in the extreme case knife-edge selection at a threshold length l_r, as in Figure 4.11. Fish below length l_r suffer no fishing mortality at all, since they are not within the range of the gear. But on reaching length l_r they are at once exposed to a high fishing mortality (as drawn in Figure 4.11 more than half of the full fishing mortality). The resultant selection curve in this case therefore starts at length l_r and rises

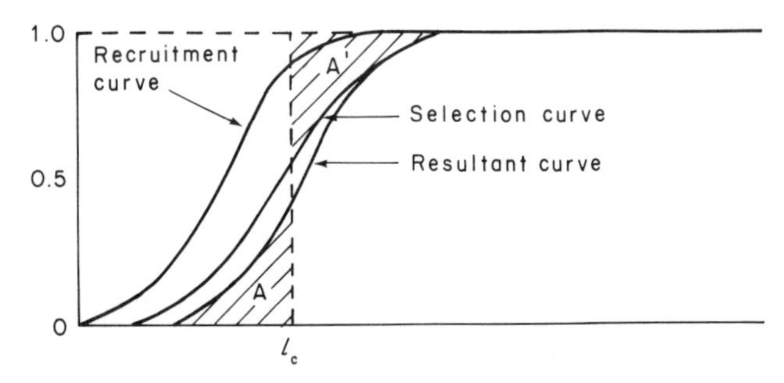

Figure 4.10 The effect of combining selection and recruitment curves

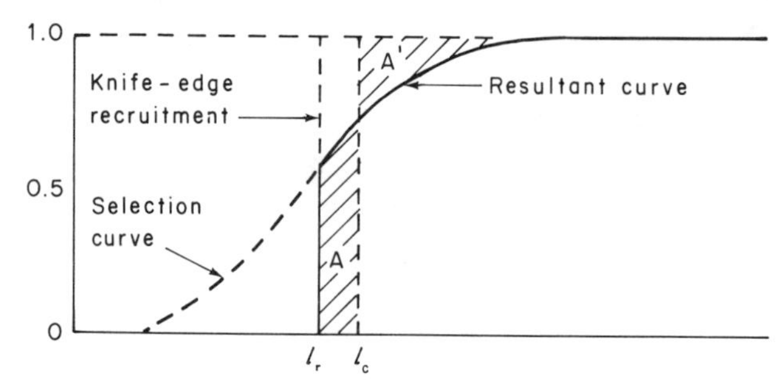

Figure 4.11 The effect of combining a selection curve and 'knife-edge' recruitment

vertically to the selection curve of the gear; thereafter it is identical with the selection curve itself. Again, however, the mean selection length l_c can be calculated from this resultant curve, in this case l_c is such that the two areas A and A′ of Figure 4.11 are equal.

The use of a gear with a certain selectivity is often combined as part of a management or conservation, with controls on the sizes of fish that can be caught, or at least landed and exposed for sale. The usual incentive for this is to improve the degree of compliance with the regulations by reducing the incentive to use, for example, a trawl with less than legal mesh size. Because selection takes place over a range, it is not possible to specify an exact size limit to compare with a specified mesh size. Every size limit greater than the lower point of the selection range will mean that an illegal net would catch slightly more legal-sized fish (at least in the selection range) than in a legal net, while any limit less than the upper point of the range would mean that a legal net would catch some undersized fish. In practice, setting the size limit toward the bottom end of the range seems preferable.

In any case it is necessary to examine the combined effect of gear selection and a size limit on the entry of fish into the catches.

If a minimum legal size limit, l_s say, is set within or above the selection range of the gear in use, the entry to the landed catch can be represented by a curve which starts at l_s and rises vertically until the true selection curve is reached (see Figure 4.12). The mean selection length l_c is calculated in a similar way and is such as to equalize the areas A and A′ of Figure 4.12.

Undersized fish will also be taken, and since they cannot be landed they must be rejected at sea. The effect of this capture and rejection on the yield from the stock depends on how many of the rejected fish survive. If all survive, the situation is the same as if the recruitment curve to the landed catch were the true selection curve.

If, on the other hand, some or all undersized fish die after rejection, each year-class is subjected to a certain fishing mortality before it reaches length l_s, as represented by the lower part of the selection curve in Figure 4.11. The resultant fishing mortality coefficient which takes account of the capture and

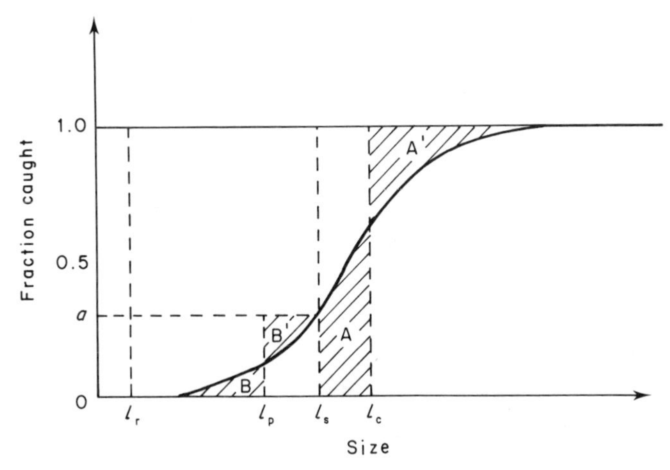

Figure 4.12 The effect of a size limit on the effective selection curve

rejection of undersized fish is obtained by calculating the mean of that part of the selection curve which is below the size limit. Calling this mean length l_p, then it is such as to equalize the areas B and B′ of Figure 4.12. In other words, it is assumed that fish are not caught at all until they reach length l_p, and are then subjected to the constant fraction a of the full fishing mortality rate until they reach size l_s (a is the ordinate at length l_s, see Figure 4.12). If growth from length l_p of length l_s occupies a period $l_s - l_p$ (see below) and the fraction q of rejected fish die, then the year-class is reduced by the factor

$$\exp[-qaF(t_s - t_p)] \tag{4.39}$$

where t_p is the age of fish of length l_p, as a result of capture and rejection of undersized fish before it enters the marketable size range. In this expression F is the full fishing coefficient to which fish are exposed when they have grown beyond the selection range of the gear.

4.4.6 Estimation of recruitment

In a previous section (4.4.1), the pre-recruit phase of the population has been compared to a black box producing a certain number of recruits into the exploited stock. Other sections have discussed how the mean age at recruitment, t_r, may be estimated. It is also desirable for many purposes, not least in understanding the working of that black box, and especially the stock/recruitment problem (see Chapter 6) to obtain estimates of the number of recruits R at age t_r. Where cohort analyses can be carried out (see section 4.3.4), this provides the most direct method, subject to the caution that the small fish just after recruitment may be subject to somewhat higher natural mortality than the exploited stock as a whole. The process of extrapolating back from the observed catches, involved in the cohort analysis procedure, may therefore underestimate the absolute number of recruits, but the relative numbers in different years will be unaffected.

Failing the detailed age–composition data required for cohort analysis, less detailed data can be used to obtain relative indices of year-class strength. For example, c.p.u.e. of a given year-class can be plotted, on a logarithmic scale, against age, and the c.p.u.e. at the age at recruitment read off from this graph.

Estimation of the absolute level of recruitment requires some measurement in absolute terms of some characteristics of the exploited stock that can be related to the strength of recruitment. For example the average numbers in the stock is given by

$$\bar{N} = R/Z$$

from which the number of recruits can be estimated as

$$R = \bar{N}Z$$

A similar, but more complicated expression can be used for biomass. Also, it may be easier to calculate yield per recruit (see Chapter 5), observe the actual yield, and thus estimate the average recruitment, than to estimate recruitment in any other way.

Besides estimating the abundance of a recruit class after the fish have been recruited, it is often valuable (e.g. in setting catch quotas in a regulated fishery) to have estimates of recruitment before the fish recruit. Though correlations have been observed between year-class strength and various environmental conditions, they have seldom if ever been either precise enough or reliable enough to provide a practical basis for prediction. This has to be based on some form of survey. Surveys of 0-recruit or pre-recruit fish have become a major element of the international programme of stock monitoring in the North-east Atlantic (Tambs-Lyche, 1978). The techniques follow the standard methods of resource surveys (see Saville 1977), though the gear used may need to be modified, e.g. by using small-meshed trawls.

REFERENCES AND READING LIST

4.2 Growth

Abramson, N. J. and P. K. Tomlinson (1961) Fitting a von Bertanlanffy growth curve by least squares, including tables of polynomials. *Fish. Bull. Calif. Dep. Fish Game*, No. 116: 1–69.

Allen, K. R. (1966) A method of fitting growth curves of the von Bertalanffy type to observed data. *J. Fish. Res. Board Can.*, **23** (2): 163–179.

Allen, K. R. (1969) Application of the von Bertalanffy growth equation to problems of fisheries management: a review. *J. Fish. Res. Board Can.*, **26** (9): 2267–2281.

Allen, R. L. (1976) A method for comparing fish growth curves. *N.Z. Mar. Freshwat. Res.*, **10** (4): 687–692.

Arnoult, J., M. L. Bauchot and J. Daget (1965) Etude de la croissance chez *Polypterus senegalus* Cuvier. *Acta Zool.*, **46**: 297–309.

Backiel, T. (1968) Ageing of coarse fish. Rome, FAO, FI:EIFAC 68/SC 1–6: 4 pp. (mimeo).
Bagenal, T. B. (ed.) (1974) *Ageing of Fish.* Proceedings of an international symposium. Old Woking, Surrey, Gresham Press, Unwin Bros., 234 pp.
Bayley, P. B. (1977) A method for finding the limits of application of the von Bertalanffy growth model and statistics estimates of the parameters. *J. Fish. Res. Board Can.*, **34** (8): 1079–1084.
Bertalanffy, L. von (1938) A quantitative theory of organic growth. *Hum. Biol.*, **10** (2):181–213.
Bhattacharya, C. G. (1967) A simple method of resolution of a distribution into Gaussian components. *Biometrics*, **23** (1): 115–135.
Blacker, R. W. (1974) Recent advances in otolith studies. In *Sea Fisheries Research*, F. R. Harden Jones (ed.). New York, Wiley, pp. 67–90.
Bourlière, F. (1946) Longévité moyenne et longévité maximum chez les vertébrés. *Ann. Biol. CIEM*, **3**: 249–270.
Brothers, E. B., C. P. Mathews and R. Lasker (1976) Daily growth increment in otoliths from larval and adult fisheries. *Fish. Bull. NOAA/NMFS*, **74** (1): 1–8.
Cassie, R. M. (1954) Some uses of probability paper in the analysis of size frequency distributions. *Aust. J. Mar. Freshwat. Res.*, **5**: 513–522.
Christensen, J. M. (1964) Burning of otoliths, a technique for age determination of soles and other fish. *J. Cons. CIEM*, **29**: 73–81.
Cloern, J. E. and F. H. Nichols (1978) A von Bertalanffy growth model with a seasonally varying coefficient. *J. Fish. Res. Board Can.*, **25** (11): 1479–1482.
Daget, J. (1952) Mémoires sur la biologie des poissons du Niger moyen. 1. Biologie et croissance des espèces du genre *Alestes. Bull. Inst. Fondam. Agr. Noire*, **14** (1): 191–225.
De Bondt, A. F. (1967) Some aspects of age and growth of fish in temperate and tropical waters. In *The Biological Basis of Freshwater Fish production*, S. D. Gerking (ed.). Oxford, Blackwell Scientific Publications, pp. 67–88.
Fulton, T. W. (1911) *The Sovereignty of the Sea.* Edinburgh.
Gallucci, V. F. and T. J. Quinn (1979) Reparameterizing, fitting and testing a simple growth model. *Trans. Am. Fish. Soc.*, **108** (1): 14–25.
Harding, J. P. (1949) The use of probability paper for the graphical analysis of polymodal frequency distributions. *J. Mar. Biol. Assoc. UK.*, **28**: 141–153.
Holden, M. J. and M. R. Vince (1973) Age validation studies on the centra of *Raja clavata* using tetracycline. *J. Cons. CIEM*, **33**: 335–339.
Holden, M. J. and D. F. S. Raitt (1974) Manual of fisheries science. Pt. 2. Methods of resource investigations and their application. *FAO Fish. Tech. Pap.*, No. 115, Rev. 1: 214 pp.
Le Guen, J. C., F. Baudin-Laurencin and C. Champagnat (1969) Croissance de l'albacore (*Thunnus albacares*) dans les régions de Pointe-Noire et de Dakar. Cah. ORSTOM (Sér.Océanogr.), **7** (1): 19–40.
MacDonald, P. O. M. and T. J. Pitcher (1979) Age-group from size-frequency data: a versatile and efficient method of analysing distribution mixtures. *J. Fish. Res. Board Can.*, **36** (8): 987–1001.
McNew, R. W. and R. C. Summerfelt (1978) Evaluation of a maximum likelihood estimator for analysis of length–frequency distributions. *Trans. Am. Fish. Soc.*, **107** (5): 730–736.
Menon, M. D. (1950) The use of bones other than otoliths in determining the age and growth rate of fishes. *J. Cons. CIEM*, **16** (3): 311–340.
Micha, J. C. (1971) Densitè de population, age et croissance du Barbeau *Barbus barbus* (L.) et de l'Ombre *Thymallus thymallus* (L.) dans l'Ourthe. *Ann. Hydrobiol.*, **2** (1): 47–68.
Moore, H. L. (1951) Estimation of age and growth of yellowfin tuna (*Neothunnus macropterus*) in Hawaiian waters by size frequencies. *Fish. Bull. USFWS*, **65**: 133–149.
Pauly, D. and N. David (1981) ELEFAN 1, a BASIC program for the objective extraction of growth parameters from length frequency data. *Meeres forsch*, **28** (4): 205–211.

Pauly, D. and G. Gaschutz (1978) A simple method of fitting oscillating length growth data with a programme for pocket calculations. ICES Doc. C.M. 1979/G. 24 (mimeo.).

Poinsard, F. and J.-P. Troadec (1966) Determination de l'age par la lecture des otoliths chez deux espèces de Sciaenidés ouest-africains (*Pseudotolithus senegalensis* C.V. et *Pseudotolithus typus* Blkr.). *J. Cons. CIEM*, **30** (3): 291–307.

Qasim, S. Z. (1973) Some implications of the problem of age and growth in marine fisheries from the Indian waters. *Indian J. Fish.*, **20** (2): 351–371.

Richards, F. J. (1959) A flexible growth curve for empirical use. *J. Exp. Biol.*, **10** (29): 290–300.

Ricker, W. E. (1969) Effects of size-selective mortality and sampling bias on estimates of growth, mortality, production and yield. *J. Fish. Res. Board Can.*, **26** (3): 409–434.

Riffenburgh, S. H. (1960) A new method for estimating parameters for the Gompertz growth curve. *J. Cons. CIEM*, **25** (3): 285–293.

Rothschild, B. J. (1967) Estimates of the growth of skipjack tuna (*Katsuwonus pelamis*) in the Hawaiian Islands. *Proc. IPFC*, **12** (2): 100–111.

Stevens, W. L. (1951) Asymptotic regression. *Biometrics*, **7** (3): 247–267.

Stott, B. (1968) Marking and tagging. In *Methods for Assessment of Fish Production in Freshwaters*, W. E. Ricker (ed.). *IBP Handbook*, No. 3: 78–92.

Tanaka, S. (1956) A method of analysing the polymodal frequency distribution and its application to the length distribution of porgy (*Taius tumifrons* T. and S.). *Bull. Tokai Reg. Fish. Res. Lab.*, No. 14: 1–15 (in Japanese). Issued also as *J. Fish. Res. Board Can.*, **19** (6): 1143–1159.

Taubert, B. D. and D. W. Coble (1977) Daily rings in otoliths of three species of *Lepomis* and *Tilapia mossambica*. *J. Fish. Res. Board Can.*, **34** (3): 332–340.

Tesch, F. W. (1968) Age and growth. In *Methods for Assessment of Fish Production in Freshwaters*, W. E. Ricker (ed.). *IBP Handbook*, No. 3: 93–123.

4.3 Mortalities

Beverton, R. J. H. and B. C. Bedford (1963) The effect on the return rate of condition of fish when tagged. *Spec. Publ. ICNAF*, No. 4: 106–116.

Beverton, R. J. H. and S. J. Holt (1956) A review of methods for estimating mortality rates in fish populations, with special reference to sources of bias in catch sampling. *Rapp. P.-V. Reun. CIEM*, **140** (1): 67–83.

Beverton, R. J. H. and S. J. Holt (1957) On the dynamics of exploited fish populations. *Fish. Invest. Minist. Agric. Fish. Food UK*. (*Series* 2), No. 19: 533 pp.

Beverton, R. J. H. and S. J. Holt (1959) A review of the lifespans and mortality rates of fish in nature and their relation to growth and other physiological characteristics. In *CIBA Colloquia on Ageing*. Vol. 5. *The Lifespan of Animals*, G. E. W. Wolstenholme and M. O'Connor (eds.) London, Churchill, pp. 142–180.

Burczynski, J. (1979) Introduction to the use of sonar systems for estimating fish biomass. *FAO Fish. Tech. Pap.*, No. 191: 89 pp.

Chapman, D. G. and D. S. Robson (1960) The analysis of a catch curve. *Biometrics*, **16**: 354–368.

Chapman, D. G., B. D. Fink and E. B. Bennett (1965) A method for estimating the rate of shedding of tags from yellowfin tuna. *Bull. I-ATTC*, **10** (5): 335–352.

Cormack, R. M. (1969) The statistics of capture–recapture methods. *Oceanogr. Mar. Biol.*, **6**: 455–506.

Doi, T. (1974) Further development of whale sighting theory. In *The Whale Problem*, W. E. Schevill (ed.). Cambridge, Mass., Harvard University Press, pp. 359–368.

Doubleday, W. G. (1976) A least squares approach to analysing catch at age data. *Res. Bull. ICNAF*, No. 12: 69–81.

English, T. S. (1964) A theoretical model for estimating the abundance of planktonic fish eggs. *Rapp. P.-V. Reun. CIEM*, **155**: 174–181.

Forbes, S. T. and O. Nakken (1972) Manual of methods for fisheries resources survey and appraisal. Pt.2. The use of acoustic instruments for fish detection and abundance estimation. *FAO Man. Fish. Sci.*, No. 5: 138 pp.

Gray, D. F. (1977) An iterative derivation of fishing and natural mortality from catch and effort data giving measures of goodness of fit. ICES Doc. C.M. 1977/F.33 (mimeo).

Gulland, J. A. (1955) On the estimation of growth and mortality in commercial fish populations. *Fish. Invest. Minist. Agric. Fish. Food UK.* (*Series* 2), **18** (9): 46 pp.

Gulland, J. A. (1963) On the analysis of double-tagging experiments. *Spec. Publ. ICNAF*, No. 4: 228–229.

Gulland, J. A. (1965) Estimation of mortality rates. Annex to Arctic Fisheries Working Group Report. Paper presented to ICES Annual Meeting, 1965 (mimeo).

Gulland, J. A. (1977) The analysis of data and development of models. In *Fish Population Dynamics*, J. A. Gulland (ed.). London, Wiley, pp. 67–95.

Jones, R. (1964) Estimating population size from commercial statistics when fishing mortality varies with age. *Rapp. P.-V. Reun. CIEM*, **155**: 210–214.

Jones, R. (1976) The use of marking data in fish population analysis. *FAO Fish. Tech. Pap.*, No. 153: 42 pp.

Jones, R. (1977) Tagging: theoretical methods and practical difficulties. In *Fish Population Dynamics*, J. A. Gulland (ed.). London, Wiley, pp. 46–66.

Jones, R. (1979) Materials and methods used in marking experiments in fishery research. *FAO Fish. Tech. Pap.*, No. 190: 134 pp.

Jones, R. (1981) The use of length composition data in fish stock assessment (with notes on VPA and cohort analysis). *FAO Fish Circ.* No. 734: 60 pp.

Murphy, G. I. (1965) A solution of the catch equation. *J. Fish. Res. Board Can.*, **22** (1): 191–202.

Pope, J. G. (1972) An investigation of the accuracy of virtual population analysis using cohort analysis. *Bull. I-ICNAF*, No. 9: 65–74.

Pope, J. G. (1977) Estimation of fishing mortality, its precision and implications for the management of fisheries. In *Fisheries Mathematics*, J. H. Steele (ed.). London, Academic Press, pp. 63–74.

Posgay, J. A. (1953) The sea scallop fishery. *Contr. Woods Hole Oceanogr. Instn* 715

Regier, H. A. and D. S. Robson (1967) Estimating population number and mortality rates. In *The Biological Basis of Freshwater Fish Production*, S. D. Gerking (ed.). Oxford, Blackwell Scientific Publications, pp. 31–66.

Ricker, W. E. (1975) Computation and interpretation of biological statistics of fish populations. *Bull. Fish. Res. Board Can.*, No. 191: 382 pp.

Ricker, W. E. (1977) The historical development. In *Fish Population Dynamics*, J. A. Gulland (ed.). London, Wiley, pp. 1–26.

Robson, D. S. and D. G. Chapman (1961) Catch curves and mortality rates. *Trans. Am. Fish. Soc.*, **90**: 181–189.

Robson, D. S. and H. A. Regier (1967) Estimation of population number and mortality rates. In *Methods for Assessment of Fish Production in Freshwaters*, W. E. Ricker (ed.). *IBP Handbook*, No. 3: 124–158.

Schumacher, A. (1971) Bestimmung des fishereilichen Sterblichkeit beim Kabeljaubestand vor Westgronland. *Ber. Dtsch. Wiss. Komm. Meeresforsch*, **21**: 248–259.

Seber, G. A. F. (1973) *The Estimation of Animal Abundance and Related Parameters.* London, Griffin, 506 pp.

Ssentongo, G. W. and P. A. Larkin (1973) Some simple methods of estimating mortality rates of exploited fish population. *J. Fish. Res. Board Can.*, **30**: 695–698.

Smith, P. E. and S. L. Richardson (1977) Standard techniques for pelagic fish egg and larva surveys. *FAO Fish. Tech. Pap.*, No. 175: 100 pp.

Welcomme, R. L. (ed.) (1975) EIFAC (European Inland Fisheries Advisory Commission) Symposium on the methodology for the survey, monitoring and appraisal of fishery resources in lakes and large rivers. *EIFAC Tech. Pap.*, (23): Suppl.1, 2 vols: 747 pp.

4.4 Selection and Recruitment

Allen, K. R. (1968) Simplification of a method of estimating recruitment rates. *J. Fish. Res. Board Can.*, **25** (12): 2701–2270.

Aoyama, T. (1961) The selective action of trawl nets and its application to the management of Japanese trawl fisheries in the East China and Yellow Seas. *Bull. Seikai Reg. Fish. Res. Lab.*, No. 23: 63 pp.

Beverton, R. J. H. (1963) Escape of fish through different parts of cod end. *Spec. Publ. ICNAF*, No. 5: 9–11.

Blaxter, J. H. S., B. B. Parrish and W. Dickson (1964) The importance of vision in the reaction of fish to driftnets and trawls. In *Modern Fishing Gear of the World*, Fishing News International and Fishing News (Books) (eds.). London, Fishing News (Books), pp. 529–536.

Boerema, L. K. (1956) Some experiments on factors influencing mesh selection in trawls. *J. Cons. CIEM*, **21** (2): 175–191.

Bowen, B. K. (1963) Preliminary report on the effectiveness of escape gaps. *Western Australia Fish. Dept. Rep.*, **2**: 1–9.

Bowen, B. K. (1971) Management of the western rock lobster (*Panulirus longpipes cygnus* George). *Proc. IPFC*, **14** (2): 139–153.

Buchanan-Wollaston, H. J. (1927) On the selective action of a trawl net, with some remarks on the selective action of drift nets. *J. Cons. CIEM*, **2** (3): 343–355.

Cassie, R. M. (1955) The escapement of small fish from trawl nets and its application to the management of the New Zealand snapper fisheries. *Fish. Res. Bull. Mar. Dep. N.Z.*, No. 11: 99 pp.

Clark, J. R. (1960) Report on selectivity of fishing gear. *Spec. Publ. ICNAF*, No. 2: 27–36.

Davis, F. M. and H. Buchanan-Wollaston (1934) Mesh experiments with trawls, 1928-33. *Fish. Invest. Minist. Agric. Fish. Food UK.* (*Series* 2), **14** (1): 56 pp.

French, R. R. (1969) Comparison of catches of Pacific salmon by gill nets, purse seines, and longlines. *Bull. INPFC*, No. 26: 13–26.

Gulland, J. A. (1956) On the selection of hake and whiting by the mesh of trawls. *J. Cons. CIEM*, **21** (3): 296–309.

Gulland, J. A. (1964) Variations in selection factors and mesh differentials. J. Cons. CIEM, **29** (2): 158–165.

Gulland, J. A. and D. Harding (1961) The selection of *Clarias mossambicus* (Peters) by nylon gill nets. *J. Cons. CIEM*, **26**: 215–222.

Hamley, J. M. (1975) Review of gill net selectivity. *J. Fish. Res. Board Can.*, **32** (11): 1943–1969.

Hamley, J. M. and H. A. Regier (1973) Direct estimates of gillnet selectivity to walleye (*Stizostedion vitreum vitreum*). *J. Fish. Res. Board Can.*, **30** (7): 817–830.

Holt, S. J. (1963) A method for determining gear selectivity and its application. *Spec. Publ. ICNAF*, No. 5: 106–151.

ICES (1971) Report of the ICES/ICNAF Working Groups on selectivity analysis. *Coop. Res. Rep. ICES* (*A*), No. 25: 144 pp.

Ishida, T. (1962) On the gill net mesh selectivity curve. *Bull. Hokkaido Reg. Fish. Res. Lab.*, No. 25: 20–25. Issued also as *Transl. Ser. Fish. Res. Board Can.*, (1338) (1969).

Jones, R. (1976) Mesh regulations in the Demersal fisheries of the South China Sea area. FAO/UNDP South China Sea Programme, Manila. Doc. SCS/76/WP/34.

Kipling, C. (1957) The effect of gill-net selection on the estimation of weight-length relationships. *J. Cons. CIEM*, **23**: 51–63.

McCombie, A. M. and F. E. J. Fry (1960) Selectivity of gill nets for lake whitefish, *Coregonus clupeaformis. Trans. Am. Fish. Soc.*, **89** (2): 176–184.

Naumov, V. M. and A. N. Smirnov (1969) Influence of the selectivity of gill nets on the quantitative and qualitative composition of commercial fish populations in the Sea of Azov. *Tr. Vses. Naucho-Issled. Inst. Morsk. Rybn. Khoz. Okeanogr.*, **67**: 262–299. Issued also as *Transl. Ser. Fish Res. Board Can.*, (1593).

Olsen, S. (1959) Mesh selection in herring gill nets. *J. Fish. Res. Board Can.*, **16** (3): 339–349.
Parrish, B. B. (1969) A review of some experimental studies of fish reactions to stationary and moving objects of relevance to fish capture processes. *FAO Fish. Rep.*, **2** (62): 233–245.
Pope, J. A. *et al.* (1975) Manual of methods for fish stock assessment. Pt 3. Selectivity of fishing gear. *FAO Fish. Tech. Pap.*, No. 41, Rev. 1: 65 pp.
Regier, H. A. and D. S. Robson (1966) Selectivity of gill nets, especially to lake whitefish. *J. Fish. Res. Board Can.*, **23** (3): 423–454.
Saville, A. (ed.) (1977) Survey methods of appraising fishery resources. *FAO Fish. Tech. Pap.*, No. 171: 76 pp.
Tambs-Lyche, H. (1978) Monitoring fish stocks. The role of ICES in the northeast Atlantic. *Mar. Policy*, **2** (2): 127–132.
Vooren, C. M. and R. F. Coombs (1977) Variations in growth, mortality and population density of snapper, *Chrysophrys auratus* (Forster), in the Hauraki Gulf, New Zealand. *Fish. Res. Bull. Fish. Res. Div. N.Z. Minist. Agric. Fish.*, No. 14: 32 pp.
Westhoff, C. J. W., J. A. Pope and R. J. H. Beverton (1962) The ICES mesh gauge. Charlottenlund Slot, ICES, 13 pp. (mimeo.).

EXERCISES

Exercise 4.1

Table 4.1 gives the length distribution of a sample of porgy (*Taius tumifrons*) taken in the East China Sea.

Table 4.1

Length (cm)	Numbers
9	509
10	2240
11	2341
12	623
13	476
14	1230
15	1439
16	921
17	448
18	512
19	719
20	673
21	445
22	341
23	310
24	228
25	168
26	140
27	114
28	64
29–30	22

What can you say about the age composition of this sample? How many age-groups can be clearly distinguished? What are their modal lengths? What are the numbers in each group? Can the later modes be more easily identified by establishing the complete frequency distribution of the early groups?

Exercise 4.2

Table 4.2 below gives the length composition of Pacific Ocean perch caught off British Columbia from 1960 to 1976 (for compactness only even-numbered years are given—data provided by W. E. Ricker and S. Westreim).

Table 4.2

Length (cm)	1960	1962	1964	1966	1968	1970	1972	1974	1976
≤ 25	3	14	1	0	8				
26	4	13	0	1	6	12	2	1	0
27	5	12	1	1	12	12	1	2	3
28	19	17	1	2	13	21	5	3	6
29	26	33	6	3	14	34	10	6	17
30	44	37	9	9	16	51	25	8	32
31	63	48	10	15	26	76	29	10	28
32	58	69	32	26	35	85	67	17	29
33	84	101	39	43	35	89	108	31	35
34	121	124	59	85	64	85	137	79	67
35	126	136	79	122	83	68	154	142	99
36	98	138	111	134	85	53	119	185	162
37	99	110	125	152	120	61	93	153	188
38	83	85	130	158	125	63	77	115	159
39	90	58	121	127	125	77	71	98	108
40	74	51	98	108	104	82	65	86	81
41	72	34	89	75	99	72	55	71	55
42	43	28	97	55	75	61	48	61	40
43	28	36	89	39	63	58	37	43	27
44	13	24	58	18	39	50	33	42	15
45	13	11	21	10	22	38	29	20	17
46	10	8	10	5	15	21	14	10	11
47	5	1	2	0	3	7	7	1	7
48	3	0	0	0	1	4	1	2	2
49	1	0	0	0	0	1	0	2	0
≥ 50	3	0	0	0	0				

Is any change or progression of modes apparent? (Plot graphically.) Calculate the average percentage frequency in each length group, and express the observed percentage in each year as a deviation from that percentage. Plot these deviations, determine progressions of exceptionally good or bad broods, and hence estimate a growth curve from 25 cm onwards. (i) How many groups of good (or bad) year-classes can be followed through? (ii) Are the length distribution in all years consistent? (iii) How many years does it take a Pacific Ocean perch to grow from 25 to 40 cm?

Exercise 4.3

The average lengths at successive ages of herring caught off Hokkaido between 1910 and 1954 are given in Table 4.3.

Table 4.3

Age	Length (cm)
3	25.70
4	28.40
5	30.15
6	31.65
7	32.85
8	33.65
9	34.44
10	34.97
11	35.56
12	36.03
13	35.93
14	37.04
15	37.70

The length–weight relationship was estimated to be given by

$$W = 0.0078 \times L^3 \text{ g}$$

Calculate the increment of growth during each year. Hence, estimate L_∞ and K graphically or by calculating the regression of increment on initial length. Repeat, using the Ford–Walford plot of l_{t+1} against l_t. Is there a difference in estimates? What account should be taken of the growth during the 13th year?

Calculate values of t_0, using equation (4.6) for different ages. Hence, estimate a mean value of t_0, and using this and the estimates of L_∞ and K, calculate the predicted value of length and weight at each age. How do these compare with the observed value?

Exercise 4.4

Posgay (1953) tagged sea scallops, and obtained data on growth during the 10 months (approximately) between tagging and recapture (Table 4.4).

Table 4.4

	Shell diameter (mm)									
Size at tagging	64	69	71	94	104	105	110	117	117	126
Size at recapture	98	102	93	115	120	126	125	127	136	138

(i) Calculate L_∞ and K.

(ii) Taking $t_0 = 0$ calculate the length at each age up to six years, and plot the growth curve.

Exercise 4.5

The size composition of samples of *Cynoglossus semifasciatus* taken off the Malabar coast of India are given in Table 4.5 (data from Seshappa, 1973):

Table 4.5

	Numbers of fish sampled				
Length (cm)	January	February	March	April	May
3 –	—	—	—	3	—
4 –	6	—	4	14	2
5 –	48	—	8	24	5
6 –	143	25	5	35	16
7 –	58	54	15	18	31
8 –	24	28	39	18	18
9 –	17	10	59	40	46
10 –	16	7	37	49	52
11 –	5	8	13	16	29
12 –	10	1	5	3	12
13 –	4	4	15	1	11
14 –	1	5	4	—	9
15 –	—	—	4	—	3

(i) What modes can be detected?

(ii) If the position of the mode in January can be neglected as being too high because of partial recruitment, and time units of one month are used, fit a growth curve to the modes in February to May. What are the values of L_∞, and K (a) using month as unit time, (b) using a year as unit time? Why might these values be unreliable?

Exercise 4.6

(a) A certain mortality causes 25 per cent of the population to die each year; what percentage of the initial population is left after two years, six months, three years? What is the corresponding instantaneous mortality coefficient? (*N.B.* Find the coefficient first.) Repeat for mortalities causing 10 per cent, 90 per cent, 50 per cent mortalities per year.

(b) Two causes of mortality, acting independently, cause mortality coefficients of 0.2 and 0.3; what is the resulting total mortality coefficient? Repeat with coefficients of 0.7 and 0.1; 1.0 and 0.3.

(c) Two causes of mortality act independently on a population; alone they would cause respectively 20 per cent and 30 per cent of the population to die within a year. Do 50 per cent of the population die within a year? If not, what is the percentage that does die? (*N.B.* Convert to instantaneous mortality coefficients.) Repeat for pairs of mortalities causing 70 per cent and 30 per cent; 80 per cent and 70 per cent mortalities.

(d) Plot y against x, on semilogarithmic paper, and $\log_e y$ against x on ordinary graph paper, for each set of y.

x	0	1	2	3	4	5	6
y_1	0.79	0.63	0.50	0.40	0.32	0.25	0.20
y_2	0.71	0.35	0.18	0.089	0.045	0.022	0.011

Do the points satisfy the equation $\log y = ax + b$ or $y = ce^{dx}$? If so, what are the values of a, b, c, d? (Estimate these from the slope and intercept of the straight-line relation.) If the y's are indices of the density of a certain year-class of fish at yearly intervals, do the data fit constant mortality coefficients? If so, what are the fractions surviving each year, and what are the instantaneous mortality coefficients?

Exercise 4.7

A research vessel in five one-hour trawl hauls caught the following numbers of fish at each age: I, 30; II, 450; III, 120; IV, 70; V, 25; VI +, 15.

A year later, the catches in 12 one-hour hauls were: I, 60; II, 960; III, 480; IV, 120; V, 72; VI +, 42.

Assuming for the purposes of this example that, with the mesh size in use, the catch per hour gives a valid index of density for two-year-old fish and older, make various estimates of the total mortality coefficient during the year. If the only data available were those for the first set of hauls, give an estimate of the avegate total mortality during the previous few years. (In practice, 5 or 12 hauls would generally be quite insufficient to provide a valid index of density.)

Exercise 4.8

A group of fish is subject in two successive years to total mortality coefficients of 0.85 and 0.80. If the number present at the beginning of the first year is 1000, what is the average number present in each of the two years? What is the total mortality coefficient estimated from these two averages?

Exercise 4.9

The average number of plaice landed at Lowestoft per 100 hours' fishing in two periods are given in Table 4.6.

Table 4.6

	Age								
Years	2	3	4	5	6	7	8	9	10
1929–38	125	1355	2352	1761	786	339	159	70	28
1950–58	98	959	1919	1670	951	548	316	180	105

Plotting numbers (on a logarithmic scale) against age:

(i) At what age are plaice fully recruited?

(ii) Calculate the average total mortality coefficients in the two periods for the fully recruited age-groups.

(iii) If the average total fishing effort on plaice in the North Sea during the two periods were, in terms of millions of hours fishing by the United Kingdom steam trawlers, 5.0 in 1929–38 and 3.1 in 1950–58, estimate from equation (4.30) the natural mortality, and the fishing mortality in the two periods.

Exercise 4.10

Table 4.7 gives the age composition of cod in terms of numbers landed per 100 ton-hours fishing by United Kingdom trawlers, in the Barents Sea, and the total fishing effort in each year.

Table 4.7

	Age										
Years	3	4	5	6	7	8	9	10	11	12 +	Effort
1932	0.02	0.38	1.49	3.16	3.81	2.16	1.55	2.20	1.48	2.80	174
1933	—	0.37	2.12	3.37	4.62	3.37	1.60	1.37	0.64	2.18	184
1934	—	0.43	3.18	6.18	5.47	8.69	2.59	0.75	0.56	1.98	164
1935	0.03	0.56	7.85	13.97	10.04	5.82	2.82	0.88	0.29	1.22	184
1936	0.1	0.58	1.75	8.61	7.96	9.28	3.50	0.93	0.19	1.13	252
1937	0.02	1.89	2.91	7.66	17.05	8.65	2.53	0.38	0.26	0.33	321
1938	0.4	1.87	7.09	5.73	9.46	9.54	4.30	1.24	0.48	0.13	253
1946	7.3	11.52	11.39	18.91	13.77	11.77	14.29	17.65	5.26	3.13	66
1947	2.1	18.25	28.42	28.42	22.57	5.99	7.10	10.45	11.01	4.74	103
1948	0.9	1.89	15.59	33.19	26.77	11.31	7.17	2.01	3.65	2.76	156
1949	4.4	12.92	18.38	34.83	20.08	9.08	6.58	2.94	1.59	2.50	171
1950	—	0.49	1.15	5.49	12.89	9.32	4.28	1.76	1.01	0.37	248
1951	0.04	1.41	8.61	13.58	11.52	6.62	4.11	1.76	1.96	0.02	313
1952	0.15	4.65	11.84	16.26	13.14	4.09	2.11	1.52	0.91	0.93	412
1953	0.02	2.02	13.77	16.41	7.20	5.49	1.62	0.70	0.15	0.25	396
1954	0.07	5.33	19.03	21.76	20.25	3.99	1.62	0.48	0.67	0.13	425
1955	—	0.73	10.83	26.82	15.70	5.44	1.93	1.30	0.57	0.19	551
1956	0.02	0.50	6.29	22.76	13.51	5.59	1.20	0.61	0.24	0.08	630
1957	0.08	1.98	3.40	9.81	9.70	6.55	1.29	0.58	0.38	0.10	457
1958	0.6	4.4	9.98	8.31	5.81	5.53	1.56	0.91	0.42	0.20	414

Tabulate the total mortality coefficient for each pair of years for each age. Examine the data and decide from what age the fish are fully represented in the catches. What evidence is there that the age at full recruitment is different in the pre- and post-war periods? Obtain a single estimate for the mortality between successive pairs of years (take the ratio 7 and older : 8 and older). By relating this mortality to the average fishing effort in the two years, estimate the natural mortality and the fishing mortality in 1958. Note that the estimate of M obtained will, as usual, include any net migration from the fished area. Note also that detailed analysis on a data set like this would normally be carried out by cohort analysis, once an estimate of natural mortality has been obtained.

Exercise 4.11

Table 4.8 (from Schumacher, 1971) gives the age composition of catches of West Greenland cod. Assuming $M = 0.2$, apply the methods of cohort analysis to the 1954 year-class to estimate the fishing mortality during each year. Repeat for other year-classes.

Table 4.8 West Greenland cod. Annual number of fish caught in each age-group (000s)

Age	1956	1957	1958	1959	1960	1961	1962	1963	1964	1965	1966
2		544	488			24		296	8	2 752	88
3	209	1 177	348	578	435	2 946	869	7 612	8 655	14 718	1 294
4	1 758	19 353	1 772	2 866	6 186	22 958	11 423	6 589	27 181	58 619	7 738
5	4 996	12 493	15 136	5 464	5 168	19 756	70 311	19 301	11 407	53 331	59 987
6	17 901	9 362	6 751	27 411	4 652	8 055	29 344	48 418	18 264	8 994	40 726
7	6 622	17 367	7 501	6 622	20 250	6 980	7 816	22 517	30 864	9 152	5 791
8	6 400	3 967	17 177	3 881	4 492	23 126	5 050	3 973	11 355	15 125	4 403
9	24 418	4 061	3 181	5 996	2 743	4 359	13 772	1 708	2 543	2 595	6 667
10	2 345	8 893	3 652	1 124	5 363	2 333	2 433	6 768	1 027	539	1 166
11	4 106	1 271	12 981	1 477	805	4 724	1 709	1 104	4 138	472	276
12	1 014	1 899	1 691	4 327	1 438	528	2 599	1 156	591	1 864	122
13	1 363	485	2 168	999	5 195	1 138	720	2 325	321	73	981
14	2 893	436	725	836	741	5 052	1 219	189	933	34	137
14^+	1 194	1 383	3 271	960	1 859	2 383	2 897	3718	747	265	234

Exercise 4.12

In 1979, 6481 male shrimps were tagged by the Kuwait Institute for Scientific Research. The numbers of returns in 10-day periods are given in Table 4.9:

Table 4.9

Days since tagging	Numbers returned
0 –	501
10 –	270
20 –	133
30 –	177
40 –	55
50 –	96
60 –	17
70 –	27
80 –	2
90 –	4

By plotting the logarithm of numbers against time estimate the total and fishing mortality coefficients. Express the coefficients in terms of 10-day periods, and as annual values.

Exercise 4.13

Shrimp (*Penaeus semisulcatus*) were tagged by the Kuwait Institute for Scientific Research in 1975 and 1976, using different colour tags (Table 4.10).

Table 4.10

Year	Colour	Number tagged	Number returned
1975	White	1325	2
	Yellow	716	29
	Red	25	0
1976	Yellow	126	2
	Red	180	0
	Blue	838	53

(i) Is the rate of return the same for all colours?
(ii) If release conditions were the same, what causes could there be for the differences?
(iii) If a correction is made for these differences, what is the maximum number of shrimp which were effectively tagged in each year?
(iv) Would your answers to (ii) and (iii) be changed if it were known that different colours were used in different places and dates?

Exercise 4.14

Garcia (personal communication) tagged shrimp in the Ivory Coast in 1973, and distinguished the returns from shrimp tagged at night, and day, with the results as given in Table 4.11.

Table 4.11

	No. tagged			No. returned		
	Day	Night	Total	Day	Night	Total
February	127	642	769	22	357	389
August	642	296	938	346	185	531
October	609	318	927	57	55	112

(i) What evidence is there that there was substantial losses at tagging?
(ii) If the water was clear in February, but turbid at other times, what may have been the primary cause of these losses? (The difference in general rate of return may be ascribed to differences in fishing patterns.)
(iii) If all shrimp tagged at night survived, what was the effective number tagged each month?

Exercise 4.15

From cod tagged in the Gulf of St Lawrence in 1955, the numbers returned by different countries in 1956 are given in Table 4.12 (from Dickie, 1963).

Table 4.12

Country	Number of tags	Total cod landings (tons)
Canada	392	62 400
France	32	28 000
Portugal	29	5 800
Spain	15	8 100

Calculate the number of tags returned per 1000 tons of fish landed by each country. What evidence is there of incomplete returns? If all tags recaptured by Canadian fishermen are returned, estimate the number actually recaptured by other countries.

Exercise 4.16

Three groups of herring tagged with internal tags were released in the North Sea in the summer of 1957. Detailed data of Danish fishing intensity in the areas around the release points were available, in terms of hours of fishing per statistical square (15 × 15 miles).

The number of tags returned, and the fishing intensity around each tagging position in the weeks following tagging are given in Table 4.13 (data adapted from ICES, 1961).

Table 4.13

Week after tagging	Liberation number					
	1		2		3	
	Tags	Intensity	Tags	Intensity	Tags	Intensity
1	21	80	80	420	3	10
2	35	64	12	60	3	5
3	30	53	10	50	8	15
4	8	26	2	10	2	10
5	10	56	1	10	—	—
Number tagged	3000		1500		3000	

Plot the tags returned per hours of fishing per square per 1000 tags released for each liberation. Hence, if the efficiency of the recovery technique is such that 90 per cent of the tags caught are returned, what is the value of q, in terms of an intensity of 100 hours' fishing per square per week?

Exercise 4.17

A plankton survey showed that 2×10^{11} eggs were laid during a spawning season. Fecundity studies showed that the averages mature female laid 10^5 eggs. Market studies showed that 3 million fish were landed during the year following, and of these 40 per cent were adult females (i.e. had spawned at least once). What percentage of the spawning females was caught during the year? If the total mortality coefficient was 1.2, what were the fishing and natural mortality coefficients?

Exercise 4.18

The percentage length–distributions of cod caught by gill-nets and purse-seines at Lofoten in 1952 are given in Table 4.14 (data from Rollefsen, 1952).

Table 4.14

Length (cm)	65 –	70 –	75 –	80 –	85 –	90 –	95 –	100 –	105 –	110 –	115 –	120 +
Purse-seine	2	4	8	13	20	28	35	34	25	15	8	5
Gill-nets	0	2	5	15	30	50	45	27	15	5	2	1

Assuming there is no selection by purse-seines, determine the form of the selection curve of the gill-nets. What size of fish is caught most efficiently by gill nets?

Exercise 4.19

(a) Cod-ends of 44 m and 112 mm fished alternatively gave catches of plaice in equal fishing times as set out in Table 4.15.

(b) A cod-end of 74 mm fitted with a small-meshed cover gave catches of whiting distributed between the cod-end and cover as in Table 4.16.

Table 4.15 Plaice

Length (cm)	44 mm mesh	112 mm mesh
10	0	0
11	1	0
12	2	0
13	6	0
14	16	0
15	18	0
16	26	0
17	64	3
18	121	4
19	182	10
20	247	13
21	292	16
22	344	37
23	367	60
24	355	132
25	276	190
26	225	177
27	147	136
28	90	97
29	57	54
30	26	33
31	24	26
32	18	13
33	10	9
34	12	7
35	5	6
36	10	5
37	6	2
38	3	4
39	2	2
40	2	3
41	0	1
42	0	2
43	0	2
44	0	1
45	0	1
46	0	1
47	0	0
48	0	0
49	0	0

Table 4.16 Whiting

Length (cm)	Cod-end	Cover
10	0	0
11	0	0
12	0	3
13	0	7
14	0	21
15	0	27
16	0	23
17	0	19
18	2	20
19	3	83
20	23	219
21	68	302
22	89	256
23	79	154
24	115	55
25	106	42
26	87	18
27	63	7
28	55	0
29	38	2
30	34	0
31	11	0
32	11	0
33	6	0
34	9	0
35	5	0
36	2	0
37	1	0
38	1	0
39	0	0

Draw the selection curves for plaice and whiting from these data and determine the mean selection lengths (l_c). Note that the number entering the net is for (a) (alternate hauls) equal to the catch in the small-meshed net, but for (b) is the sum of the fish required in the cod-end, and those which went through the meshes, i.e. in the cover.
Calculate selection factors.

Exercise 4.20

Vooren and Coombs (1977) gave data on the length composition of each age-group, and the selection of different trawl mesh sizes, for the snapper fishery of the Hauraki Gulf, New Zealand (Table 4.17).

Table 4.17

Fork length (cm)	Age (years)[a]						Mesh size (cm)[b]		
	2	3	4	5	6	7	8.0	10.0	11.4
15	5						95	100	100
16	—						87	100	100
17	5						74	100	100
18	10						55	100	100
19	5						36	98	100
20	3						19	93	100
21	5						9	84	100
22	5	5					3	69	99
23	9	9	1				1	50	96
24	5	33	2	1			0	31	90
25	22	22	9	3			0	16	77
26	17	12	11	8			0	7	60
27	9	9	19	3			0	2	40
28	—	7	19	16	13		0	0	22
29	—	2	22	27	27	18	0	0	10
30	—	1	9	13	20	18	0	0	4
> 30			8	29	40	64	0	0	0

[a]Percentage of age-group in stated length-class.
[b]Percentage of fish in given length-class released by trawl of stated mesh size:

If the fishing mortality on fully recruited fish is 1.0, calculate the fishing mortality on each age from two to seven, for each of the three mesh sizes.

CHAPTER 5

Yield-per-recruit

5.1 WHY CALCULATE YIELD-PER-RECRUIT?

Chapter 4 has described how estimates can be obtained of the basic parameters of the exploited population. The present chapter describes part of the process of using these estimates to advise policy-makers on the immediate and long-term effects of different actions. Particular attention is given to the effects of alteration of the two parameters that can be directly controlled—the amount of fishing, as measured by the fishing mortality F, and the way the fishing is distributed on different sizes of fish, as measured by the age at first capture, t_c.

It should be stressed to begin with that any fish stock is part of a complex natural system. It is therefore very difficult to state with any certainty what the effects of any action will be. Some of the problems are discussed in later chapters, which examine the extent to which changes in fishing pressure on one stock may affect catches from other stocks, and, as regards events within one stock, the question of the effects of changes in adult stock abundance on the average level of subsequent recruitment.

The scientist, and those he is advising, must therefore accept the fact that a comprehensive assessment of the long-term effects of any pattern of fishing is difficult, and is likely to be subject to inaccuracies. The situation is considerably brighter when considering the fate of a brood or year-class of fish once they have been recruited to fishery. Many of the uncertainties, for example the influence of adult stock size or environment on recruitment are past. Growth rate and natural mortality may still change, but after recruitment these changes are comparatively small. Once reasonable estimates of growth and natural mortality are attained, then the task of calculating the yield from a given recruiting year-class is straightforward, requiring no further assumptions, and the result is not subject to variation because of environmental conditions, etc. Therefore the scientist can give a clear and unequivocal answer to questions such as 'What pattern of fishing will give the greatest yield from the year-class of fish that has just been recruited?', as well as providing other information on, for example, the egg production from the year-class during its life, under different patterns of fishing. These answers are clearly important in themselves, but they also provide a solid foundation on which advice can be given, and tentative conclusions reached, regarding the longer-term effects of different patterns of fishing. Therefore, calculation of yield from a given recruitment—usually expressed as yield-per-recruit—is a basic element in the assessment of any fish stock for which the data allow the calculations to be made. Once these calculations have been made, it is possible to make various assumptions about the stock/recruitment relation (see section 6.2) in order to make assessments of the likely long-term effects of different patterns of fishing.

There are other reasons for calculating yield-per-recruit, or providing advice in terms of yield-per-recruit. In many stocks recruitment is highly variable, and the variations seem to be independent of adult stock, being determined entirely by environmental factors. For these stocks the predictions of future catches beyond the life span of the year-classes already recruited is very difficult, and it is impossible to say that, for example, if a larger mesh size is used, the catches in some future year will be some specified percentage higher than the present—recruitment to the stock in that year may be very poor. What can be said is that, with a larger mesh size, catches in any future year will be some percentage higher than they would have been if the mesh size had not been increased. This conclusion is well expressed by the advice that the yield-per-recruit will be increased by some given percentage.

5.2 CALCULATION OF YIELD-PER-RECRUIT

Consider the fate of number R of recruits entering the fishery at some age t_r, and let us define:

N_t = number of fish alive at age t

M = natural mortality coefficient, supposed constant

F_t = fishing mortality coefficient, which may vary with age

W_t = average weight of an individual fish of age t

Then, in the short interval $t, t + \mathrm{d}t$, the numbers $\mathrm{d}C_t$, and weight $\mathrm{d}Y_t$ which are caught will be given by

$$\mathrm{d}C_t = F_t N_t \mathrm{d}t$$
$$\mathrm{d}Y_t = F_t N_t W_t \mathrm{d}t$$

and hence the total catch in numbers and weight will be obtained by summing, or integrating, over the whole period that the group of fish is exposed to the fishery, say from the age of recruitment t_r up to some limiting age t_L.

$$C = \int_{t_r}^{t_L} \mathrm{d}C_t = \int_{t_r}^{t_L} F_t N_t \mathrm{d}t \tag{5.1}$$

$$Y = \int_{t_r}^{t_L} \mathrm{d}Y_t = \int_{t_r}^{t_L} F_t N_t W_t \mathrm{d}t \tag{5.2}$$

Two steps are now required: first to obtain expressions for F_t, N_t, and W_t (bearing in mind that N_t will be a function of the fishing mortality when the fish were younger than t), and then making the calculations. This may be done, as suggested above, by integration, or by simple addition of the catches taken during successive periods, e.g. years or months of life. Historically, the latter was the first to be used, but in the pre-computer age could be used only in a qualitative or descriptive way. For short and therefore numerous periods, the computations were too lengthy, while for longer periods there are no single and accurate expressions for the average weight or number caught (see Exercise 5.1).

To carry out the integration, the terms on the right-hand side of equations (5.1) and (5.2) have to be expressed in terms that are suitable for integration. If the integration is to be reasonably simple, the mortality coefficient can only vary in extremely simple ways. Usually the natural mortality is assumed constant, and the fishing mortality zero up to the age at first capture, t_c, and thereafter constant, i.e.

$$F_t = 0; \qquad t \leq t_c$$

$$F_t = F = \text{constant}; \qquad t > t_c$$

From this the value of N_t can be deduced from equation (4.17); since

$$Z_t = M; \qquad t \leq t_c$$

$$Z_t = F + M; \qquad t > t_c$$

$$N_t = R\exp[-M(t - t_r)]; \qquad t \leq t_c$$

where R = number of fish alive at time $t = t_r$, the age at recruitment, and

$$N_t = R'\exp[-(F+M)(t - t_c)]; \qquad t > t_c$$

where R' = number of fish alive at time $t = t_c$, and $R' = R\exp[-M(t_c - t_r)]$. The numbers caught can be calculated from equation (5.1) as

$$C = \int_{t_c}^{t_L} R'F\exp[-(F+M)(t - t_c)]\,\mathrm{d}t$$

$$= R'\frac{F}{F+M}(1 - \exp[-(F+M)(t_L - t_c)])$$

or

$$C = R\frac{F}{F+M}\exp[-M(t_c - t_r)](1 - \exp[-(F+M)(t_L - t_c)]) \qquad (5.3)$$

or ignoring the last term, if t_L is large

$$C = R\frac{F}{F+M}\exp[-M(t_c - t_r)]$$

or

$$C = \frac{F}{F+M}R' \qquad (5.4)$$

The most convenient form for the expression of weight for integration is the exponential $W_t = W_1 \exp[G(t - t_1)]$. This is adequate for describing the growth during a limited period, say from time t_1 to t_2; hence the yield in weight during period t_1 to t_2 can be given by

$$Y = \int_{t_1}^{t_2} FR' \exp[-(F+M)(t-t_c)]\, W_1 \exp[G(t-t_1)]\,dt$$

$$= FR' \exp[-(F+M)(t_1-t_c)] W_1 \int_{t_1}^{t_2} \exp[-(F+M-G)(t-t_1)]\,dt$$

$$Y = FR'W_1 \exp[-(F+M)(t_1-t_c)] \frac{1}{F+M-G} \times (1-\exp[-(F+M-G)(t_2-t_1)]) \tag{5.5}$$

While this is satisfactory for moderately short periods, it is little use for long periods—the growth equation predicts that a fish will become infinitely large at great ages. The equation for weight at age that combines adequate realism and relative convenience when integrating is the von Bertalanffy (see section 4.2), i.e.

$$W_t = W_\infty \{1 - \exp[-K(t-t_0)]\}^3$$

To integrate, the right-hand side can be expanded and expressed as the sum of four terms, i.e.

$$W_t = W_\infty \sum_{n=0}^{3} U_n \exp[-nK(t-t_0)]$$

where $U_0 = 1$, $U_1 = -3$, $U_2 = 3$, and $U_3 = -1$.

Then,

$$Y = \int_{t_c}^{t_L} F \cdot R'W_\infty \exp[-(F+M)(t-t_c)] \sum_0^3 U_n \exp[-nK(t-t_0)]\,dt$$

or, writing $t - t_0 = (t - t_c) - (t_c - t_0)$ and rearranging the terms

$$Y = FR'W_\infty \sum_0^3 U_n \int_{t_c}^{t_L} \exp[-(F+M+nK)(t-t_c)] \exp[-nK(t_c-t_0)]\,dt$$

On integrating this becomes

$$Y = FR'W_\infty \sum_0^3 \frac{U_n}{F+M+nK} \exp[-nK(t_c-t_0)] \times [1-\exp[-(F+M+nK)(t_L-t_c)]] \tag{5.6}$$

Again, this expression may be simplified if t_L is sufficiently large for the last term to be neglected. Then the yield is given by

$$Y = FR'W_\infty \sum_0^3 \frac{U_n \exp[-nK(t_c - t_0)]}{F + M + nK}$$

or, substituting for R′

$$Y = FR\exp[-M(t_c - t_r)]\,W_\infty \sum \frac{U_n \exp[-nK(t_c - t_0)]}{F + M + nK} \tag{5.7}$$

Besides the total yield, some other quantities are important. Thus the yield is F times the average biomass of fish in the exploited phase, B', and therefore

$$B' = R'W_\infty \sum \frac{U_n \exp[-nK(t_c - t_0)]}{F + M + nK} \tag{5.8}$$

Similarly, the average number in the exploited phase is, from equation (5.4), obtained by dividing the catch in numbers by F:

$$N' = \frac{R'}{F + M} = \frac{R\exp[-M(t_c - t_r)]}{(F + M)}$$

The average weight of the individual fish in the exploited phase is therefore

$$\bar{W} = \frac{B'}{N'} = \frac{Y}{C} = W_\infty \sum_0^3 \frac{U_n(F + M)\exp[-nK(t_c - t_0)]}{F + M + nK}$$

The above equations are normally expressed as yield per recruit (or numbers, biomass, etc.), i.e.

$$Y/R = F\exp[-M(t_c - t_r)]\,W_\infty \sum \frac{U_n \exp[-nK(t_c - t_0)]}{F + M + nK} \tag{5.9}$$

A number of computer programs exist for this, and similar expressions.

Although these equations contain several parameters, several of these, for instance W_∞, only appear as a constant factor. Since, in considering yield-per-recruit (or average biomass per recruit, etc.), we are principally concerned with changes, or relative yield-per-recruit at different levels of fishing mortality or age at first capture, these constant factors do not matter.

As shown mathematically below, the yield-per-recruit can, apart from constant factors, be described by three parameters, one a characteristic of the fish, and the others of the fishery. The first is the ratio of the growth parameter K, and the natural mortality M. This determines chances of a fish completing much of its potential growth before dying of natural mortality. If M/K is small, these chances are good, the stock (in the absence of fishing) will contain many relatively large fish, and it will pay, in terms of getting the best yield from a recruit level, to fish relatively lightly, and with a high size at first capture. If M/K is large, many fish will die before completing much of their growth, and it will therefore pay to fish relatively hard and with a low size at first capture so as to catch the fish before they die of natural causes. It may be noted that, as suggested in section 4.3.5, M/K appears to be relatively constant within a group of fish species. In the absence of data for a given species this constancy may help in making reasonable assumptions from data on related species.

The two characteristics of the fishery are the amount of fishing, conveniently expressed as the ratio of fishing to natural mortality (F/M), or the exploitation ratio, E, $(=F/(F+M))$, and the relative size at first capture, conveniently expressed as $c = l_c/L_\infty$ $[=1-\exp\{-K(t_c-t_0)\}]$.

Hence, we can write

$$\exp[-K(t_c-t_0)] = 1-c$$

and

$$\begin{aligned}\exp[-M(t_c-t_r)] &= \exp[-M(t_0-t_r)]\exp[-M(t_c-t_0)]\\ &= \exp[-M(t_0-t_r)](1-c)^{M/K}\end{aligned}$$

and equation (5.6) can be re-written as

$$Y = FRW_\infty \exp[-M(t_0-t_r)](1-c)^{M/K}\sum\frac{U_n(1-c)^n}{F+M+nK}$$

or, taking out constant terms characteristic of the stock but independent of F and c, and writing

$$Y = Y'[RW_\infty \exp[-M(t_0-t_r)]] \tag{5.10}$$

and

$$Y' = E(1-c)^{M/K}\sum\frac{U_n(1-c)^n}{1+\dfrac{nK}{M}(1-E)} \tag{5.11}$$

The expression on the right-hand side of equation (5.10) contains only E, c, and M/K. The values have been tabulated by Beverton and Holt (1959) for a series of values of M/K from 0.25 to 5.00. For each value of M/K the tables can be used directly as a yield–isopleth diagram, giving a number proportional to the yield-per-recruit, as a function of c (i.e. size at first capture), and of E (a measure of the intensity of fishing). From this the effects of different actions on the fishery can be easily determined, though some care should be taken in interpreting the effects of changes in the amount of fishing at high values of E. This is because if E is large (greater than about 0.5), quite small changes in E correspond to large changes in F. For purposes other than quick identification of whether the present value of E (or F) is greater or less than the value giving the greatest yield-per-recruit, it is desirable to plot the entries in the table (i.e. the indices of Y/R), explicitly as a function of F, and also to calculate and plot other important measures of the state of the fishery, e.g. Y'/F, which will be an index of the catch per unit effort (c.p.u.e.) (assuming constant recruitment).

There are a number of ways in which the results of calculations of yield-per-recruit can be expressed. Since their main purpose is to provide advice on the likely results of changes from the present situation—for example, as a

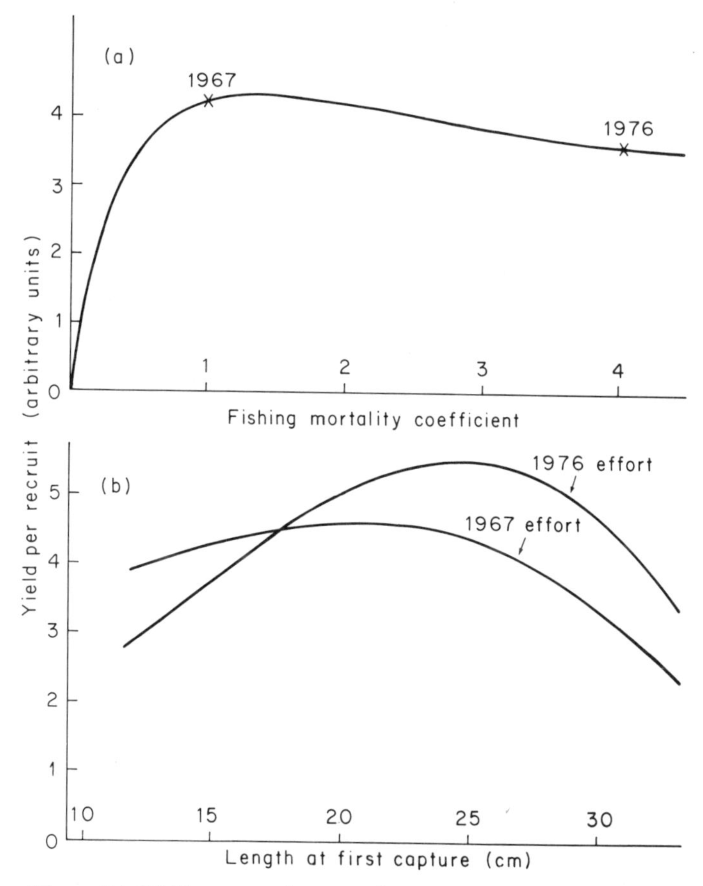

Figure 5.1 Yield per recruit curves for *Saurida undosquamis* in the Gulf of Thailand. Showing (a), above, the yield as a function of fishery mortality (crosses mark the mortalities in 1967 and 1976), and (b), below, the yield as a function of size at first capture (or mesh size), for the values of fishing mortality occurring in 1967 and 1976

result of a regulation of mesh size, investment in more vessels, etc.—this is most obviously done by one or other of two figures. The yield-per-recruit can be expressed either as a function of fishing mortality, F, keeping the age at first capture to constant, or as a function of t_c, keeping F constant. In each case the present value of F, or t_c respectively, shall be clearly indicated. For the latter curve, it will be helpful in practical interpretation when dealing with fisheries in which the control of t_c will be achieved by control of mesh size, to indicate on the x-axis the mesh size equivalent to each size at first capture (see Figures 5.1a and b).

These figures do not give the whole story. In particular they do not show the effect of simultaneous changes in both F and t_c, which can be important. For example, increasing the mesh sizes may reduce the importance of reducing

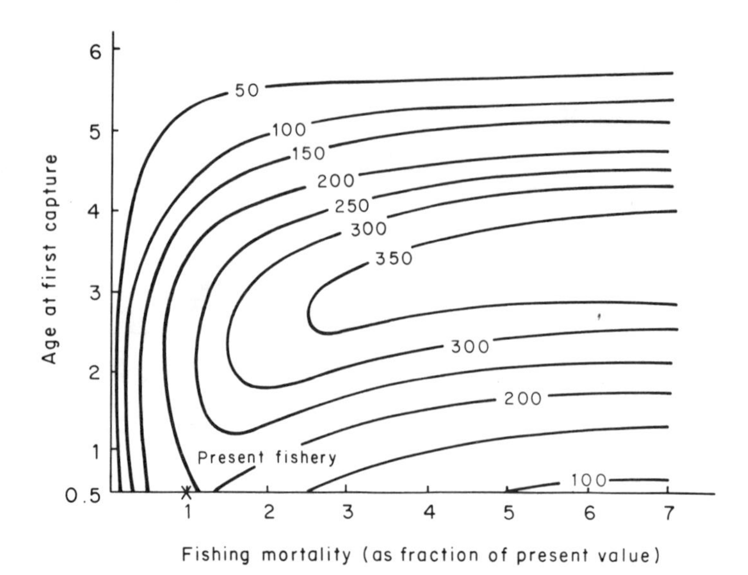

Figure 5.2 Yield–isopleth diagram for the croaker (*Pseudotolithus typus*) in Congolese waters. The figure shows lines of constant yield per recruit as a function of fishing mortality and age at first capture. (Data from unpublished MS by A. Fontana.)

the amount of fishing on a heavily fished stock; on the other hand, for a moderately heavily fished stock an increase in fishing (due, for example, to additional investment) may make an increase in mesh size more desirable. To deal with this situation an isopleth diagram giving contours of equal values of yield-per-recruit as functions of the values of t_c and F is useful (see Figure 5.2). The lines corresponding to the present values of t_c and F should be indicated. The curves of Figures 5.1a and b then correspond to sections parallel to the x and y-axes respectively.

The shortcoming of these figures is that they show what will actually happen only if the parameters (values of M, F, etc.) have been correctly estimated. This is most unlikely, and the true effect on the yield-per-recruit of say, increasing the amount of fishing by 10 per cent, will not be exactly as indicated in Figure 5.1a. The advice should therefore contain some information on the likely results if the parameters are not exactly as estimated. It is not possible to take account, in any single presentation, of all possible variations in the parameters. In practice, estimation errors for different parameters are not equally large or equally important. Usually, at least when the age of individual fish can be readily determined, the growth parameters can be estimated with good precision. Estimation of the effective age at first capture, t_c, can be readily made if adequate samples from the commercial catches are available. The difficulties of estimation mostly concern mortalities, and especially the separation of natural and fishing mortality. Good samples of age composition can yield estimates of total mortality, Z, but it requires something extra—data over a period in which

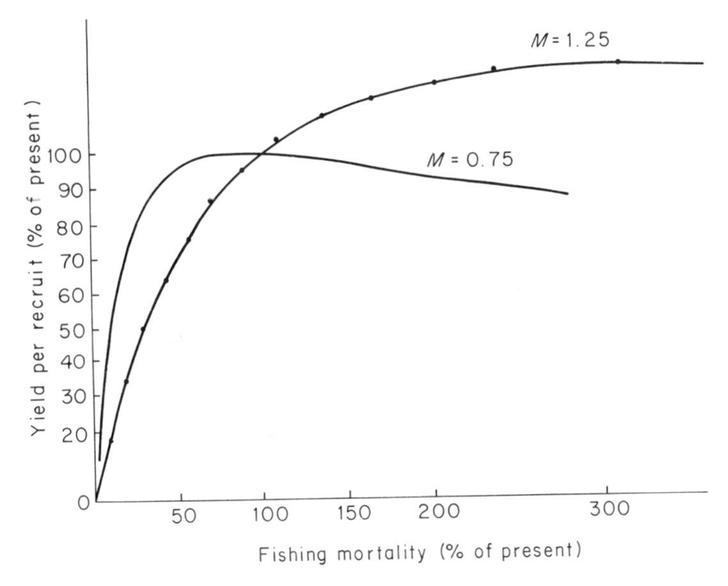

Figure 5.3 Yield-per-recruit (as per cent of current value) as a function of fishing mortality (also as a percentage of current value) for two values of natural mortality. (Data based on values for *Caranx ruber* round Jamaica—$L_\infty = 51, l_c = 17\,\text{cm}, K = 0.25, Z = 2.0$)

effort has varied considerably, successful tagging experiments, etc.—to estimate F and M separately. The results of yield-per-recruit calculations therefore need to be presented using different assumptions concerning the value of M. Presentation of the expected results of changes in t_c (or of the minimum mesh size used) gives rise to no difficulties; the results for different values of M can be plotted in the same figure, preferably showing the yield-per-recruit for each value of t_c as a percentage of that taken with the present value of t_c. In the case of mortalities, the usual situation is that the current value of total mortality, Z, is reasonably well estimated. Different assumptions about the value of M therefore imply differences in the estimate of the current value of F. For the purposes of advising on the effects of changing the amount of fishing, the results are best shown by plotting the yield-per-recruit as a percentage of the current value, as a function of the fishing mortality, also expressed as a percentage of the current value (see Figure 5.3).

The other approach dealing with the integrations of equations (5.1) and (5.2) is to consider separately the yield from a series of short periods, i.e. to write

$$C = \sum_i C_i$$

$$Y = \sum_i Y_i$$

where C_i, Y_i = catches in numbers and weight during the time interval t_i, t_{i+1}, and the first interval is counted from t_r, i.e. $t_1 = t_r$.

The great advantage of this approach is that it allows greater flexibility in the variation of parameters with time. In particular, account can be taken of seasonal variations, especially in growth (or weight-at-age), and of variations in fishing mortality other than a simple pattern of recruitment or selection. In particular, it is possible to consider the effects of two or more fisheries on the same stock, but exploiting different parts of the stocks, for example one on spawning fish and another on the feeding concentration.

The simplest way of taking account of complications in growth which are not dealt with by growth curves (e.g. the von Bertalanffy equation) fitted to the whole exploited life span, is to use a single mean weight $\bar{W}_i$ for each interval. This weight can be taken directly from actual observations at sea or on fish markets without reference to any theoretical growth curve.

This assumes that the weight of fish alive during an interval will not be affected by the pattern of fishing. For short intervals this may be true with a sufficient degree of precision, but when the interval is long enough for growth to be appreciable (especially at low mortalities), a more accurate description is needed. The most convenient expression to use in combination with mortality rates is the exponential. This has been used when deriving equation (5.5) for the yield over the whole exploited life span. For such a long period the exponential does not fit very well, but is more satisfactory for shorter periods. We can then write

$$W_x = W_i \exp(G_i x)$$

where W_i = weight at age t_i, W_x = weight at age $t_i + x$, where $0 < x < t_{i+1} - t_i$, and G_i is chosen to fit the growth during the ith period, i.e. to make

$$W_{i+1} = W_i \exp[G_i(t_{i+1} - t_i)]$$

Complexities in mortalities are best dealt with by assuming that within any interval they are constant, but can vary between intervals. That is, in the case of two fisheries exerting fishing mortalities F', F'', Z_t, the mortality at any time $t, t_i < t < t_{i+1}$ is given by

$$Z_t = M_i + F_i' + F_i'' \tag{5.12}$$

This formulation is somewhat too general, in that the number of combinations of possible fishing mortalities that could be examined is enormous. They need to be limited in order to make any comprehensible analysis, and to provide meaningful advice to decision-makers. One obvious simplification is to assume that the pattern of each fishery is unchanged, i.e. it exerts the same relative mortality on fish in each age interval. That is, to write

$$F_i' = q_i' f' \tag{5.13}$$

$$F_i'' = q_i'' f'' \tag{5.14}$$

where q_i', q_i'' are fixed, being determined by the characteristics of the fishery, and the concern is with the effects of changing f' or f'', the fishing efforts in the two fisheries. In this case, once the sets of values of M_i, q_i', and q_i'' have been determined, the calculations can be restricted to two variables, f' and f''.

To make these calculations, we have for the catch in numbers by the first fishery

$$C_i' = \int_{t_i}^{t_{i+1}} F_i' N_i \exp[-(M_i + F_i' + F_i'')(t - t_i)]\mathrm{d}t$$

hence

$$C_i' = \frac{F_i' N_i}{F_i' + F_i'' + M_i}[1 - \exp\{-(M_i + F_i' + F_i'')(t_{i+1} - t_i)\}]$$

and for the catch in weight, using two alternative formulations of the weight of the individual

$$Y_i' = \int_{t_i}^{t_{i+1}} F_i' \bar{W}_i N_i \exp[-(M_i + F_i' + F_i'')(t - t_i)]\mathrm{d}t$$

or

$$Y_i' = \int_{t_i}^{t_{i+1}} F_i' W_i \exp[G_i(t - t_i)] N_i \exp[-(M_i + F_i' + F_i'')(t - t_i)]\mathrm{d}t$$

i.e.

$$Y_i' = \frac{F_i' N_i \bar{W}_i}{Z_i}[1 - \exp\{-Z_i(t_{i+1} - t_i)\}] \tag{5.15}$$

$$Y_i' = \frac{F_i' N_i \bar{W}_i}{Z_i - G_i}[1 - \exp\{-(Z_i - G_i)(t_{i+1} - t_i)\}] \tag{5.16}$$

Also

$$N_i = N_{i-1} \exp[-Z_{i-1}(t_i - t_{i-1})] \qquad i \geq 2 \tag{5.17}$$

and $N_1 = N_t = R$, the number of recruits.

Similar equations can be written for Y_i''. The method can obviously be easily extended to more than two fisheries, each with its own pattern of fishing mortality at different ages. Equations (5.17), plus either equations (5.15) or (5.16), together with a set of input values of M_i, F_i and $\bar{W}_i$ (or G_i plus an initial value of W_1), define a set of simple but numerous calculations. These can readily be carried out by computer. It is not the intention of this manual to discuss computer methods in detail, since these details will vary with the type of computer available. The programming involved is in any case not very complicated, if each set of values of F_i is treated separately. With a little more complex program, the values of F_i' and F_i'' can be determined from equations (5.13) and (5.14). In that case the values of q_i' (and q_i'' and similarly for more than two fisheries) are treated as constant input values, and the program can be rerun for sets of different values of f_i' and f_i''.

The results of these last calculations can be particularly easily presented to decision-makers in the form of $n+1$ graphs (where n is the number of independent fisheries). Of these graphs, n should show the effect on yield-per-recruit of altering the value of f for that fishery, the fishing effort in all other fisheries being kept constant. For ease of comparison it is useful to plot Y/R as a function of the effort in the fishery being varied, as a percentage of the current effort. For the $n+1$th graph, it can be taken that the effort in all fisheries alters in a similar manner (i.e. f'/f'' remains constant), and the yield-per-recruit is again plotted as a function of effort, expressed as a percentage of the current effort.

5.3 CHANGES IN SELECTION

In favourable circumstances—full supply of the necessary data, and fisheries in which the assumptions made are realistic—the methods of the previous sections can be applied to assess the effect of changes in selectivity of the gear in use. When circumstances are less favourable—and this can occur in many fisheries—it is still possible to make assessments of selectivity changes, provided sufficient information is available about the sizes of fish caught. This is because a change in selection involves a simple trade-off between catching a certain number of small fish (those retained by the present gear, but which will be released by the more selective gear), and catching a rather smaller number of bigger fish (that proportion of the fish released that ultimately get caught, after they have had time to grow).

The simplest and perhaps the most important question concerning the effect of a possible change in selectivity is whether the change will result in an increase or decrease in the average catch. This can be done, following Allen (1953), by comparing the weight of the individual fish of the size it is proposed to release, with the probability of their being caught again later and their average weight when caught. That is, if:

W_c = weight of individual fish at the mean selection length

$\bar{W}$ = mean weight of fish in the catch larger than W_c

then the catch will be increased by releasing the fish if $E\bar{W} > W_c$, where E = probability of a released fish being caught later; this will be equal to $F/(F+M)$ if the fishing and natural mortalities are constant. Or, in terms of the ratio of fishing to total mortality, the catch will be increased if $E > W_c/\bar{W}$. The right-hand side of this inequality can be readily calculated from simple observations of the size composition of the catches.

Some more detailed effects of changes in selectivity can also be determined from a knowledge of the size composition of the landings.

Following Gulland (1961, 1964) the calculations can be done in two stages: first, the immediate effect can be estimated by taking account of the observed length–composition and the selectivities of the gears used before and after the change in selection. This will give the catches immediately after the change, and also the numbers of fish released. Second, the long-term effect is estimated by considering the proportional increase in catches caused by the additional catches from these released fish, which will occur after they have had time to grow.

Mathematically, immediately following a change in gear, the numbers of any given size, l, of fish landed by the new gear (e.g. increased mesh size) will be given by

$$ {}_lN_{\mathrm{k}} = {}_lN_1 \frac{{}_lr_2}{{}_lr_1} \tag{5.18}$$

where ${}_lN_1$ = numbers landed by old gear, ${}_lN_{\mathrm{k}}$ = numbers landed by new gear, ${}_lr_1$ = proportion retained by old gear, and ${}_lr_2$ = proportion retained by new gear.

Also, if ${}_lW$ = average weight of fish of length l then the total weight W_1, caught with the old gear, is given by

$$ W_1 = \sum {}_lW_l N_1 \tag{5.19}$$

and the weight caught with the new gear immediately after the change W_{k} will be given by

$$ W_{\mathrm{k}} = \sum {}_lW_l N_{\mathrm{k}} \tag{5.20}$$

Also, the total numbers landed by the old gear, N_1, and retained by the new gear, N_{k}, can be written

$$ N_1 = \sum {}_lN_1 \quad \text{and} \quad N_{\mathrm{k}} = \sum {}_lN_{\mathrm{k}} $$

If the average weights ${}_lW$ and the relevant numbers are arranged as parallel columns, these calculations of sums of products are simple to carry out and many of the more advanced pocket calculators can be readily programmed to do the computation. The immediate change in landings can be expressed as a proportion, L, of the initial landings, where

$$ L = \frac{W_1 - W_{\mathrm{k}}}{W_1} $$

When an increase in mean selection length takes place the fish released by the new gear, but which could have been caught by the old gear, will grow, and after a time reach the size at which they will be retained by the new gear, and the catches will increase.

The number, N_{R}, of fish released, which would previously have been landed, will be equal to $N_1 - N_{\mathrm{k}}$. If quantities of small fish are caught and discarded at sea, and not landed, the total numbers released will be greater than the reduction in the landings, and may be expressed as

$$ N_{\mathrm{R}} = \sum_l {}_lN_{\mathrm{c}} \left(1 - \frac{{}_{\mathrm{r}}l_2}{{}_{\mathrm{r}}l_1} \right) \tag{5.21}$$

where ${}_lN_{\mathrm{c}}$ are the total numbers of fish of length l caught by the old gear including discards. Usually the data on the number and size of fish discarded will not be good enough to use equation (5.21) directly. Instead there may be some estimates on the proportion of the catch that is discarded (more often in terms of weight rather than numbers). From this the total number discarded may be estimated. If the change in selectivity is large (e.g. a large increase in mesh size) it may be reasonable to assume that all these (presumably very small) fish will be released. For smaller changes some intermediate proportion released will have to be assumed. In either case the number of fish previously discarded (less any that might be able to survive the catch-and-discard process) which will be released by the new gear should be added to the number of releases calculated from the length composition of the landings to give the best estimate of N_R, the total number released.

Of these fish a number $N'_R = N_R e^{-MT}$ will survive to ${}_2t_c$, the mean age of first capture with the new mesh, where T is the difference between ${}_2t_c$ and the average age of the fish released. In turn, of these fish a number EN'_R will ultimately be caught, where E is the exploitation ratio ($= F/(F + M)$ when the mortalities are constant). As all the mortalities, etc. for fish greater than ${}_2l_c$ are unchanged, the size distribution of these fish will be the same as that of the N_k fish, so that after a period the catches with the new gear will be increased by a proportion Q, when

$$Q = \frac{EN'_R}{N_k} \tag{5.22}$$

Thus the long-term landings, W_2, will be given by

$$W_2 = W_1(1 - L)(1 + Q) \tag{5.23}$$

This level of landings may not be reached until several years after the change in selectivity. It will only be complete after a period, equal to the life span of the fish in the fishery, has elapsed since the change in selectivity, although it will be close to this value at a rather earlier date. The catches in the interim period can be estimated as follows. In order to estimate the landings at time t', suppose that fish of the size released will have grown to a certain length l'. Then, at a time t' after the change in mesh size, the fish larger than l' will have been fished by the old mesh when they were of a size in the selection range, so that the stock and catches of these fish will be nearly unaffected by the change, and the weight caught of fish length $l > l'$ would be ${}_lW_lN_R$. Fish smaller than length l' would have been under the influence of the new mesh during their life in the exploited phase so the stock and catches of these sizes could be the same as in the long-term equilibrium state with the new mesh, and the weight caught of fish length $l < l'$ would be $(1 + Q)_l W_l N_R$. The total weight caught will be W', given by

$$W' = \sum_{0}^{l'} (1 + Q)_l W_l N_k + \sum_{l'}^{\infty} {}_lW_lN_k$$

i.e.

$$W' = Q \sum_{0}^{l'} {}_lW_lN_k + W_k \tag{5.24}$$

The method can be extended to stocks fished with more than one gear, some of which may not change their selectivities. Here equations (5.18) and (5.20) are applied separately to each fleet to find the immediate effects. The factor Q will be the same for all, and is given from the extension of equation (5.22) in the form

$$Q = \frac{E \sum N'_R}{\sum N_k} \tag{5.25}$$

where the summation of N_k includes the fish caught by all gears (omitting any fish, caught by gears whose selectivity does not change, smaller than the new selection length), but the summation of N_R will, of course, only include gears whose selectivity changes. The calculations necessary to assess the results of a mesh change when two gears, possibly catching different sizes of fish, are operating are set out in the work sheet below.

Work sheet for the assessment of changes in selectivity (for two gears with different size compositions of their catches)

PART 1 Immediate effects

(A)	(B)	(*C*)	(*D*)	(E)	(F)	(G)	(H)	(F)	(G)	(H)
				$\frac{(D)}{(C)}$		$(F)\times(1-(E))$	$(F)\times(E)$		$(F)\times(1-(E))$	$(F)\times(E)$
		Proportion retained			Numbers for gear A			Numbers for gear B		
l	$_lW$	$_lr_1$	$_lr_2$	$\frac{_lr_1}{_lr_2}$	$_lN_1$	$_lN_R$	$_lN_K$	$_lN_1$	$_lN_R$	$_lN_K$

$\sum {_lN}$

$\sum {_lN} \times {_lW}$

$$L = \frac{W_R}{W_1}$$

PART 2 Long-term effects

N_R: Number released
Gear A
Gear B
Discards: ———
Total: ———

N_K: Number in catches immediately after change
Gear A
Gear B
Non-regulated gears: ———
Total: ———

$\frac{N_R}{N_K} =$ ———

$T =$

$MT =$

$e^{-MT} =$

$\frac{N_R}{N_K} e^{-MT} =$

			Gear A		Gear B	
E	Q	$1+Q$	$(1-L)(1+Q)$	G	$(1-K)(1+Q)$	G

$$Q = E\frac{N_R}{N_K}e^{-MT} \qquad G = 1-(1-L)(1+Q)$$

REFERENCES AND READING LIST

Beverton, R. J. H. and S. J. Holt (1959) Tables of yield functions for fishery assessment. *FAO Fish. Tech. Pap.*, No. 38: 49 pp.

Caddy, J. F. (1977) Approaches to a simplified yield per recruit model for crustacea, with particular reference to the American lobster, *Homarus americanus. Manuscr. Rep. Ser. Mar. Sci. Dir. Can.*, No. 1445: 14 pp.

Francis, R. C. (1974) TUNPØP, a computer simulation model of the yellowfin tuna population and the surface fishery of the eastern Pacific Ocean. *Bull. I-ATTC*, **16** (3): 233–279.

Francis, R. C. (1977) TUNPØP, a simulation of the dynamics and structure of the yellowfin tuna stock and surface fishery of the eastern Pacific Ocean. *Bull. I-ATTC*, **17** (4): 215–269.

Gulland, J. A. (1961) The estimation of the effect on catches of changes in gear selectivity. *J. Cons. CIEM*, **26** (2): 204–214.

Gulland, J. A. (1964) A note on the interim effects on catches of changes in gear selectivity. *J. Cons. CIEM*, **29** (1): 61–64.

EXERCISES

Exercise 5.1

A certain species of fish recruits to the fishery at a weight of 200 g; at six-month intervals thereafter it weighs, on the average, 0.4, 0.9, 1.5, 2.3, 3.1, 3.7, 4.1, 4.6, 4.9, and 5.1 kg. During each six-month period, it can be assumed that 10 per cent of the fish alive at the beginning of the period die from natural causes. What will be the catch in weight during the first five years in the fishery, from a brood of fish numbering 1 million at the time of recruitment if during each six-month period fishing takes (a) 5 per cent of the fish alive at the beginning of the period, (b) 15 per cent and (c) 40 per cent. [Assume, for the purpose of the exercise, that natural mortality always accounts for 10 per cent of the fish alive at the beginning of the period. Is this a reasonable assumption? What is a more likely figure for the percentage dying of natural mortality during each six-month period at the highest rate of fishing?] [Assume also that the mean weight of fish during each period is the average of the weights at the beginning and end of each period.]

Exercise 5.2

The growth curve for North Sea plaice is given approximately by

$$l_t = 68.5(1 - e^{-0.1(t+0.8)})$$

where l_t is the length in centimeter. The young fish live along the coast and do not move into the main offshore fishery grounds until, on the average, an age of 3.7 years. The natural mortality is estimated to be 0.1.

Given that the mesh size currently used is small enough to retain all the fish that have recruited to the main fishing grounds, calculate l_c, the mean length at first capture, and $c(=l_c/L_\infty)$. Determine the form of the relation of the yield-per-recruit as a function of the fishing mortality.

(i) If the current value of F is 0.3, what would be the expected change in yield-per-recruit (as a percentage of the present value) as a result of (a) an increase in the fishing effort by 33 per cent, and (b) a reduction in fishing effort by 50 per cent?
(ii) What would be the expected change in c.p.u.e.? [Assume recruitment does not change, and calculate $1/F \times Y/R$.]
(iii) What change in fishing mortality would be required to produce the greatest yield-per-recruit with the present age at first capture?

Assuming that the fishing mortality remains at its present value ($F = 0.3$), examine the form of the relation of yield-per-recruit to size at first capture.

(iv) What size at first capture will give the greatest yield-per-recruit? If the selection factor for trawls is 2.2, what mesh size would give the greatest yield-per-recruit? What percentage increase in yield-per-recruit would result from the use of this mesh size?

Examine the combined effects of changes in the amount of fishing and age at first capture. (Plot as an isopleth diagram.)

Exercise 5.3

The growth curve of North Sea haddock is given by $l_t = 55 \times [1 - e^{-0.25(t+0.2)}]$, and the effective age at recruitment is one year. The selection factor is 3.2, and the present mesh size is 75 mm. The total mortality under current conditions has been estimated, with good precision, to be 1.0; the natural mortality is believed to be between 0.1 and 0.3.

Using convenient values of M/K, examine the form of the relation between Y/R and F for different values of M. What is the effect of increasing values of M on the general form of the curves? Also examine the effect of different values of M on the form of the curve of Y/R as a function of t_c.

For each value of M used, calculate the present value of F, and the tabulated value of Y/R. Hence, for each value of M, express Y/R for other values of F as a percentage of the current value.

For $M/K = 0.5$, 0.75, 1.0, and 1.5:

(a) What would be the percentage change in Y/R from increasing F by 50 per cent?
(b) What would be the percentage change in Y/R from decreasing F by 30 per cent?
(c) What percentage change in F would be required to give the greatest yield-per-recruit?
(d) If the fishing mortality is maintained at the present value, what size at first capture will give the greatest yield-per-recruit? What is the corresponding mesh size?

Exercise 5.4

Data for the United Kingdom trawl fishery for haddock at the Faroe Islands are given in Table 5.1, showing the numbers landed in each 5 cm length group, their average weight, and the proportion of each group retained by a 90 mm and a 120 mm mesh.

(a) If the present mesh is 90 mm, what would be the percentage reduction in catch (in weight) immediately following an increase to 120 mm?
(b) If it takes an average of six months for the fish released to grow to the size of first

Table 5.1

Length (cm)	Average weight (g)	Percentage (90 mm)	retained (120 mm)	Numbers landed (000s)
25–29	160	50	3	8
30–34	260	80	16	314
35–39	420	98	41	1084
40–44	630	100	72	1409
45–49	880	100	94	1370
50–54	1190	100	100	952
55–59	1570	100	100	465
60–64	1760	100	100	255
65–69	2530	100	100	124
70–74	3100	100	100	73
75–79	3800	100	100	30
80 +	4600	100	100	9

capture of the 120 mm mesh, and that the natural mortality of fish in the selection size is 0.2, and $E(=F/(F+M))=0.5$, how many of the fish released by a 120 mm mesh will ultimately be caught? If there were no other fishery on haddock, what would be the long-term effect of changing the mesh size? [Hint, first calculate the gross increase in the catches immediately following the change.]

(c) What would be the long-term effects on both fleets (as percentages, and actual weights) if there was also a line fishery operating, taking an average annual catch of 5 million fish weighing 6000 tons, and $E=0.7$? Assume all line-caught fish are larger than the selection size of the 120 mm mesh.

(d) What would be the effects if the trawlers discarded 30 per cent of the catch (by numbers), and these fish were all small enough to escape through a 120 mm mesh? (Let $T=1$ year, say.)

Exercise 5.5

About 150,000 tons of oil sardine are currently taken, on the average (though there is a large year-to-year fluctuation) with a variety of inshore gears by small-scale fisheries off the Indian west coast.

One set of population parameters that has been used is as follows: $L_\infty=210$ mm, $K=0.6$, $t_0=-0.3$, $Z=1.5$, and $M=0.4$.

(a) What other values of M might be guessed at, e.g. by comparison with K, see suggestion in section 4.3.7?

(b) Examine the form of the yield-per-recruit curves, assuming that the mean size at first capture is 10.5 cm.

(c) A suggestion has been made to build a number of medium-sized purse-seines, and it has been calculated that they would catch, on present catch rates, 75 000 tons in an average year.

For values of M of 0.4, 0.6, 0.8 and 1.2:

(i) Calculate the value of F implied by this estimated catch rate (estimate as 50 per cent of F in existing fisheries).

(ii) Assuming average recruitment is not affected, what would the value be of F exerted by the total fishery, the total catch in an average year, and the catch by the existing fishermen, and by the new boats?

CHAPTER 6

Variation in the parameters

6.1 THE RELATION OF VARIATION TO OTHER EVENTS

In the previous chapters it has been assumed that the basic natural parameters of the fish population—the growth pattern of the individual fish, the natural mortality and recruitment—remain constant. This is obviously not true; in a year of good food supply, growth is likely to be unusually good, while in unfavourable years natural mortality may be high. The same event may have different effects on interrelated species. For example, in one year disease caused a high mortality in herring in the Gulf of St Lawrence, sick herring were easily caught by cod, and the latter had an unusually high growth in that year. Above all, recruitment can be highly variable from year to year even when the size and condition of the spawning state seems to be remaining about constant.

These variations can be of considerable practical and scientific interest in themselves. Variations in year-class strength can greatly affect catches in the years immediately following the recruitment of a particularly good or bad year-class, and it would clearly be useful for the fishing industry to have predictions of these fluctuations. For the purpose of this manual the important question is the degree to which changes in the parameters can invalidate assessments of the effect of different policies (e.g. increasing the amount of fishing) which have been made on the assumption that the parameters are constant. As in the case of variations in the catchability coefficient, q (see section 2.3), much depends on the reason for the variation.

Three types of variations can be distinguished: those due entirely to natural environmental changes; those related to changes in the abundance of the stock being considered (and hence, indirectly linked to the amount of fishing in the stock); and those related to changes in other stocks (and to fishing on these stocks). The last of these is discussed further in Chapter 7, which deals with evaluation of the whole ecosystem (or at least of substantial parts of it, rather than looking at individual species in isolation), and considers the interaction between species. Detailed discussion of the first type is beyond the scope of this manual, but the possible influences of, especially, long-term trends due to climatic changes should always be borne in mind. The most careful assessment has little practical value if the stock virtually disappears because of climatic changes. The review of Cushing and Dickson (1976) provides a convenient entry to the literature on this topic.

The main concern of the present chapter is thus with the second type—density-dependent changes, especially in recruitment. However, the effects of other sources of variation, which can to some extent be considered as acting randomly with respect to the fish stock and the fishery, cannot be ignored. The practical implications of various policies, and the preferences between them may be quite different when random variations in, for example, year-class strength are taken into account, as compared with policies based on constant parameters, even when the average values are the same as those used in the constant parameter analysis. For example, fishermen may be almost as concerned with the year-to-year variation in their catches (and especially how well or badly they do in the particularly bad year), as in the average catch over a period (see exercise 6.1). Certainly a fishery administrator cannot consider two policies as being equivalent, one of which results in catches of 5500 tons each year, and the other in 1000 tons in one year and 10 000 tons in the next, even though the average catch is the same. In this case the constant catch would be preferred, but the choice would be more difficult if the second strategy gave catches of 3000 tons and 12 000 tons in alternate years.

In such a situation it is not the job of the stock assessment scientists to determine which policy is best. His main job is to provide enough information on average catches and their expected variance so that a rational choice can be made—possibly by feeding the assessment data into a more broadly based economic model of the fishery. His job does, however, go a little beyond this. Decision-makers tend to prefer simple answers and simple advice, and therefore would like to have just the information on average catches. The stock assessment expert should therefore consider whether this indeed is likely to give adequate information on which to base decisions, and if it seems inadequate, to be sure that the additional information on the distribution about the average is also properly spelled out. In particular, he may need to consider the questions of relative risk and, for example, how the chances of a collapse of the stock under unexpected bad environmental conditions might be changed as a result of different policies.

6.2 DENSITY-DEPENDENT GROWTH AND MORTALITY

Any of the natural parameters of a single species discussed in Chapter 4 can change as a result of changes in population density, but it is convenient to treat changes in recruitment (R) separately from and in more detail than changes in either growth (K, L_∞, or similar parameters for other growth equations) or natural mortality (M). This is because they tend to act in different ways. Here, R is likely to decrease as stock decreases, and hence reinforce and exaggerate changes in stock. Growth values are likely to decrease and natural mortality increase as the stock increases, thus counterbalancing and damping out changes. The practical importance of changes in recruitment, therefore, is likely to be much greater. Scientific advice based on the assumption of constant parameters is unlikely to be badly wrong (and the direction of predicted changes is likely to be correct) if in fact natural mortality or growth changes as a result of changes in stock density (and thus indirectly of changes in fishing policy), but serious errors are likely to be made if the possibility of changes in recruitment are ignored.

In separating changes in recruitment and in growth or mortality, the distinction should not be overemphasized. The mechanics whereby recruitment is maintained at lower levels of spawning stock involve changes in growth or mortality among the eggs, larvae, or young fish, and the distinction made here is essentially between these changes in the pre-recruit phase (which cannot be studied by the traditional tools of stock assessment of commercial fisheries), and those involving changes in the exploited phase of the population.

Changes in natural mortality in the latter phase are difficult to study. Usually M will be estimated indirectly as the difference between total mortality, Z, and fishing mortality, F, and most of the techniques used (e.g. regressing Z on fishing effort f) assume a constant M, or at best provide an estimate of the average value of M over a period. Point estimates of M for particular years are extremely difficult to obtain, and it is therefore correspondingly difficult to estimate changes in M, and relate these to changes in stock density. Some of the factors causing natural mortality can, however, be studied. The incidence of disease, or the abundance and distribution of predators, can be observed, and if this appears related to density, then some relation of density to M can be deduced, if only in a qualitative way. A more quantitative approach to predation is also possible when studying the interaction between two species.

In practice, density-dependent changes in M tend to be ignored when assessing the studies of a single stock. This is probably safe, but it is better to at least check what errors might be involved in doing so. This check can easily be done by repeating yield-per-recruit calculations for different values of F or of age at first capture, and instead of using a constant value of M, using one that decreases with increasing F or decreasing age at first capture. By examining the effects of different ranges of M (e.g. either M changing from 0.2 to 0.19 or from 0.2 to 0.1 as the value of F is doubled), and comparing these with the values for $M = \text{constant}$, it will be possible to tell how large the density-dependent effect would have to be to affect the analysis seriously, and hence the advice given.

Density-dependent growth is much easier to study, provided that fish can be aged by otoliths, etc. From a series of years in each of which the average size of fish at each age, as well as the stock abundance, is known, it is relatively straightforward to relate growth to density, though care should be taken to relate changes in growth to the correct time and density. In long-lived fish changes in growth during the first year or two can result in differences in length at age for many later years, even if the growth in these years is normal ('normal' in this sense should refer to the expected annual increment for a given size, rather than for a given age).

The first step in studying density-dependent growth is therefore to estimate the growth of fish of different ages and sizes during a particular year. In some circumstances the entire growth history of an individual fish can be read from its scales. Then, for example, if a 36 cm fish is caught in 1978 it may be possible not only to determine from the number of rings on the scale that it has completed six years growth, but from the width of each ring that its length at the end of each growth year was 18.1, 27.3, 29.7, 33.2, 34.1, and 35.3 cm; thus, it grew from 29.7 to 33.2 cm in 1975. Given a set of data like this it is clearly possible to describe the growth in 1975, and relate it to the density in that year. As noted in section 4.2.3 there are a number of possible theoretical growth curves that can be used. For the present purpose that of von Bertanlanffy is convenient and can be used in the form of equation (4.4), with $T = 1$. With few data each fish can supply one individual data point; alternatively the average length of all fish when completing three years growth at the beginning of 1975, and when completing four years growth in the beginning of 1976 can be estimated, and a mean growth for all three-year-old fish in 1975 calculated. This will give a data point for each age-group sampled.

A growth curve may be fitted to these data immediately, but this will result in two parameters (K and L_∞) for each year. While these can be related to the density in the year concerned, it can be awkward, unless the data are extensive, to handle two variables. It will be easier and should still result in an adequate description of how growth varies with density, to assume that one parameter, e.g. K, is constant (equal, for example, to the value already determined from fitting growth curves to all available data without taking into account density-dependent effects), and estimating a value of L_∞ for each year. There are some theoretical reasons for supposing that the value of L_∞ is more likely to vary as a result of density changes, K remaining constant (Beverton and Holt, 1957). Whether or not this is true does not affect the advantages of a greater simplicity of handling only one variable. Having calculated L_∞ for each year, the relation between it and the density can be examined, for example by regressing L_∞ on B (the biomass) as estimated, for example, from the catch per unit effort. (c.p.u.e.). For more accurate work it may be better to estimate the biomass from the numbers at each age, as derived from cohort or similar analysis (see section 4.3.4) and the average weight of each age-group during the year in question. That is, B_i, the biomass in a given year (e.g. at the beginning of that year) will be given by

$$B_i = \sum W_{ij} N_{ij}$$

where N_{ij} = number of fish age j at beginning of year i, and W_{ij} = average weight of fish age j at beginning of year i. As written, the limits of the summation have not been defined. Other things being equal, they will normally be from the age at recruitment, t_r, up to the oldest fish present. Pre-recruit fish are thus omitted from the analysis. This is probably reasonable because these fish often behave differently (e.g. are in different areas) from post-recruits, and because data on pre-recruit abundance are often difficult to obtain. However, some of the more obvious density-dependent changes in growth have occurred in these stages; also, among the recruited fish there are often differences in food or distribution, such that changes in density may affect growth in some ages more than others.

In particular, changes in fishing policy will affect the abundance in the older fish more than that of younger fish. If we are interested in seeing how density-dependent growth might affect the constant parameter estimates of, say, how the yield-per-recruit would be increased by a certain change in fishing effort, we should look first at the effects among the older fish. That is, the value of L_∞ should be calculated separately over different age-groups, say from three to six years, and the biomass calculated for the same ages, i.e. from the expression

$$B_{3-6} = \sum_{3}^{6} W_{ij} N_{ij}$$

Once the possibility of density-dependent growth has been examined, the incorporation of this effect into yield assessments is moderately straightforward. An immediate and explicit formulation of the yield to be expected from any combination of fishing mortality or age at first capture is not easy because the formulae as set out in section 5.2 require values of the growth parameters (e.g. K and L_∞), and these will depend on values of the biomass (or other measure of abundance) which will be among the outputs of these calculations. (It may be noted here that since some of the calculations and logical arguments depend on absolute abundance, rather than indices, it has to be implicitly assumed for these calculations that recruitment is constant, and the estimates will be of yield, and not yield per recruit.) Given sufficient algebraical ingenuity it is possible to derive explicit formula to deal with this problem (e.g. Beverton and Holt, 1957, section 18), but given the ability to calculate values of yield, biomass, etc. quickly for different sets of parameters other ways are simpler. One is for each pattern of fishing to guess a likely value of biomass B; from this to estimate L_∞ and K, and use these to estimate, from the yield equation of section 5.2, values of yield, etc. The output from these calculations will include a new estimate for the biomass B'. If B and B' are the same, or not very different, all is well. Otherwise a second calculation should be done, using a better input value of biomass, e.g. $B'' = \frac{1}{2}(B + B')$.

6.3 DENSITY-DEPENDENT RECRUITMENT

The question of the degree to which recruitment depends on the density or abundance (not necessarily the same thing)—usually referred to as the stock–recruitment problem—is currently one of the more interesting and more important problems of fishery research. For recent studies on stock and recruitment see Parrish (1973) and Cushing (1977). It is scientifically interesting because the stock–recruitment process seems to lie at the heart of the control of fish populations, and to be the main mechanism in determining how populations maintain themselves at around the level they do, rather than continuously declining or expanding. It is of great practical importance because the failure to take account of the possible effect of a decline in adult stock on subsequent recruitment, and thus to take measures to maintain the spawning stock, can lead to a collapse of the stock. The history of several major pelagic fish stocks (California sardine, North Sea herring) illustrate the importance of the stock–recruitment problem.

Despite this, the most obvious feature of the pattern of recruitment to most fish stocks is that it bears no obvious relation to the abundance (or other characteristic) of the parent stock. Rather, it is clear that the strength of the year-class arising from the spawning in a particular year is determined mostly by environmental factors at some early stage (or stages) in the life of that year-class; e.g. temperature just after spawning, food supply at the time of absorption of the yolk-sac, etc.

At the same time, it is equally obvious that the strength of the recruitment cannot be completely independent of the abundance of the parent stock. In fact the parent stock will determine a distribution of possible recruitments; the particular value within this distribution will be determined by the various environmental factors. The stock–recruitment problem can then be considered as the problem of determining whether or not the distribution of possible strengths of recruitment (viewed either as a simple statistical probability, or as a function of various environmental factors) is the same for all sizes of adult stocks—or at least all sizes likely to be met in practice. The mean of the distribution (i.e. the average recruitment corresponding to a given adult stock) is obviously a most important characteristic, but other features can also be of practical importance. For example, it has been suggested that, while there may be little differences under 'normal' environmental conditions, a small adult stock may be made less able than a large one to produce a reasonable recruitment under poor conditions.

The principal objective, so far as the assessment of the stock and advice for management is concerned, is to determine the relation between the abundance of the adult stock and the average strength of the subsequent recruitment. This objective is not entirely sufficient and, so far as possible, needs to be expanded to provide a description of the distribution of different recruitment strengths at each level of stock abundance. This is not easy. There are two simple methods of looking at the data: to plot recruitment against parent stock and look for a possible regression, or to separate the data into two or three groups according to the size of the adult stock (e.g. above or below the median size), and to compare the frequency distribution of the recruitment in each group. The problem is that the scatter is nearly always great, and the number of points small—only one new observation can be made each year, and for only a few fisheries do detailed data go back for more than a decade. The results are therefore nearly always inconclusive. Here a statistical caution should be emphasized; there is a very big difference between being unable to show (to any given level of statistical significance) that a certain relation exists, and being able to show that no significant (in a practical, not a statistical sense) relation exists. In particular, if there is no statistically significant correlation between stock and recruitment, this does not imply that the recruitment is necessarily independent of adult stock. Also, as pointed out by Woltes and Ludwig (1981), the size of the spawning stock is seldom known accurately. This can change the appearance of the curve, so that recruitment may appear independent of adult stock. A simple way of showing this is to consider also the relation between the survival from eggs to recruitment (as measured by the ratio R/S), and the adult stock S. If indeed average recruitment is the same for all values of S, then survival must decline exactly fast enough to balance an increase in stock, and there should be a clear inverse correlation between survival and stock. If, on analysing the data, there is indeed a significant negative correlation between R/S and S (and none between R and S), this strengthens the evidence that average recruitment does not change much with changes in adult stock. It will probably leave open the question whether there is *some* change in average, R, and whether this change could be sufficient to affect the assessment of the long-term effects of different rates of fishing. More often the correlation between S and R/S will also be non-significant, showing that, on this analysis at least, nothing definite can be said about the relation between stock and recruitment.

Another way of looking at the possible significance of the observed pairs of stock and recruitment is to consider the slope of the linear regression of recruitment on parent stock, and its confidence limits. These limits can be calculated by ordinary statistical procedures. The upper limit should be looked at most carefully. Ignoring for the present the fact that a linear regression is most unlikely to describe the stock–recruitment relation over the whole range of stock sizes, the line corresponding to this upper confidence limit (i.e. with the greatest slope that is not unlikely) shows how greatly average recruitment could be affected by changes in adult stock. If this effect is large enough to make a serious decline in abundance possible under fishing (see section 6.3), then more study is obviously needed, and also the managers of the fishery should be advised to be cautious. The lower confidence limit is of less importance, and is mainly interesting, when compared with the upper limit, in showing how much uncertainty surrounds the stock–recruitment relation.

6.3.1 Fitting stock–recruitment curves

One difficulty in determining an empirical relation between observed pairs of adult stock and subsequent recruitment is that the true relation between adult stock and the average value of recruitment is not simple. It must pass through the origin; for small stocks may be approximately proportional (though the question of 'depensatory' mortality has been raised, see Clark (1974), which if it occurs could cause an S-shaped bend towards the origin), but flattens out at higher stocks and may even decrease—in fact resembles a typical yield curve.

Two basic theoretical curves exist. Both are derived from the basic relations

$$\frac{dN_t}{dt} = -Z_t N_t \tag{6.1}$$

where N_t = number of eggs, larvae or young (pre-recruit) fish alive at time t, after spawning, Z_t = mortality coefficient, and

$$N_0 = kS; \qquad N_T = R$$

where S = abundance of the spawning stock, R = number of recruits, T = age of recruitment, and k = measure of the fecundity of the stock.

Though both Z and T may vary with density, it is simpler (and probably not misleading) to incorporate the density effects into Z. In doing this the mortality may be considered as a function of current density of the young, N_t, or of the abundance of the spawning stock, S. Taking the simplest, i.e. linear, density-dependent effect, we have

$$Z_t = b_t + c_t N_t \tag{6.2}$$

or

$$Z_t = b_t' + c_t' S \tag{6.3}$$

where, b_t, c_t, b_t', c_t' are independent of density, but will in general vary with age.

Using equations (6.2) and (6.3) and doing the necessary algebra gives the following relations, derived by Beverton and Holt (1957) and Ricker (1954), respectively; the notation differs slightly from their usages

$$R = \frac{S}{A + BS} \tag{6.5}$$

$$R = CSe^{-DS} \tag{6.6}$$

where A, B, C, D are constant.

The first equation gives a curve that rises to an asymptote, while the second reaches a maximum (where $S = 1/D, R = C/De$) and decreases at higher stocks. The latter curve certainly describes the situation in some stocks (notably salmon) better; it is also somewhat more convenient to handle. It will therefore be used in the rest of this section, but it must be remembered that the other equation, and indeed a number of other equations that can be produced with a little algebraical ingenuity, also exist, and have valid, but different, theoretical justifications.

To fit equation (6.6) to observed data, it can be rewritten in the following form, putting subscripts for the observed values in a given year:

$$R_i/S_i = Ce^{-DS_i}$$

or

$$\log_e(R_i/S_i) = \log C - DS_i \tag{6.7}$$

or, noting that S will rarely be known, some index B_i of adult biomass (e.g. c.p.u.e.) will have to be used. Simplifying the constants, this gives

$$\log(R_i/B_i) = a - bB_i \tag{6.8}$$

This is in a linear form, relating $\log R/S$ to S, and can therefore be fitted by simple regression techniques.

It may be noted that fitting requires the estimation of two unknown parameters (C and D or a and b); this is the same as that involved in calculating the ordinary linear regression of recruitment on stock, or testing the correlation between stock and recruitment. While, therefore, this more realistic curve makes no greater demands on data, it makes no less. If, as is common, the number of pairs of observation of stock and recruitment are small, or recruitment variable, the result of the fitting procedure is likely to leave considerable uncertainty regarding the actual relation. This uncertainty may be somewhat reduced by bringing in additional information.

One is to note that in equation (6.6) the term CS relates to total numbers, or absolute abundance, whereas the term DS, concerning the density-dependent effect, relates to density. The distinction will be meaningful if there is any change in the area inhabited by the stock, for example due to changes in environmental conditions. If, therefore, there is evidence that the extent of the distribution of the stock (or more particularly of the spawning stock, and of the pre-recruit fish) varies, and there is reasonable measures of the area in each year ($= A_i$), then equation (6.7) can be rewritten as

$$\log_e(R_i/S_i) = a - \frac{bS_i}{A_i} \tag{6.9}$$

Alternatively, following Csirke (1980), we can note that the c.p.u.e. (U_i) normally measures density, rather than abundance, and if another estimate of abundance is available, say from cohort analysis, or acoustic survey, equation (6.7) can also be written as

$$\log_e(R_i/B_i) = a - bU_i \tag{6.10}$$

A different approach is to use the observation of Cushing (1971), that considering data from a large number of stocks, there was some systematic variation in the degree of density dependence as measured by the value of the coefficient b in the equation

$$R = kP^b \tag{6.11}$$

It may be noted that this equation, which only passes through the origin if $b > 0$ does not give a satisfactory description over all values of P, but does give a useful measure of the shape of the stock–recruitment curve over the range of observed values. If $b \simeq 0$, R is almost constant; for $0 < b < 1$, the curve exhibits an increasing degree of direct density dependence, i.e. corresponds to the ascending limb of a general curve; while if $b < 0$, recruitment decreases with adult stock, corresponding to the right-hand decreasing limb.

Cushing found a good correlation between the value of b determined from equation (6.11) and the fecundity—more exactly there was a good linear regression (with negative slope) of b on the cube root of the fecundity of the fish. This suggests an approach whereby the value of b is estimated from the observed fecundity, and regression of Cushing. This means that only a single parameter (k of equation 6.11) has to be estimated from the observed data of stock and recruitment.

6.4 RECRUITMENT AND ENVIRONMENT

Since the major part of the variation in recruitment cannot be accounted for by changes in adult stock, environmental factors of one kind and another must be responsible. There is therefore a considerable volume of studies attempting to relate recruitment to one or other of these factors. These attempts have faced the same difficulties as those met in trying to determine the relation between recruitment and the abundance of the adult stock—the small number of observations and the large variation. The problem is in fact much harder. There is little choice of how to estimate the adult stock, but there is a wide choice of possible environmental parameters (temperature of the water, wind strength, abundance of different types of food or predator, etc.), and these may be measured at the time of spawning, or at some other time during the first few months of life of the new year-class. It is therefore dangerously easy to find some set of data which provides a high level of correlation with past recruitments. Unfortunately, few of the many correlations that have been observed (and in several cases published) have stood the test of still providing a good correlation in the years subsequent to the time of the study.

This is no reason why the relation between year-class strength and environment should not be studied, but it is a reason for treating correlations arising from such study with caution. They should be as far as possible supported with biological evidence of the mechanisms involved, and tested against observations in later years.

So far as assessments are concerned, one great potential value of establishing the quantitative effects of environmental factors is that it should remove much of the scatter about the relation between recruitment and adult stock. This should enable this relation to be determined with greater confidence. That is, studies on recruitment should not consider the effects of only adult stock, or only environment, but should as far as possible treat them together (e.g. Parrish and MacCall, 1978).

6.5 DENSITY-DEPENDENT RECRUITMENT AND TOTAL YIELD

Section 6.4 has discussed the problem of density dependence, and ways of studying the relation between stock and recruitment. This study should always be done, despite the fact that the results of the study are in practice often highly inconclusive, because of the importance of the relation to an understanding of the dynamics of a fish stock, and its ability to withstand disturbances, whether due to fishing or natural variations, and to the provision of advice on management.

This importance is best illustrated by Figure 6.1. For the purposes of visualizing what happened, it is convenient to consider a stock with a single generation, where the fish reproduce at one year old and then die. The more realistic situation of several generations present at once, and repeated breeding, does not affect the argument, and makes only a slight difference to the pattern of changes after a disturbance.

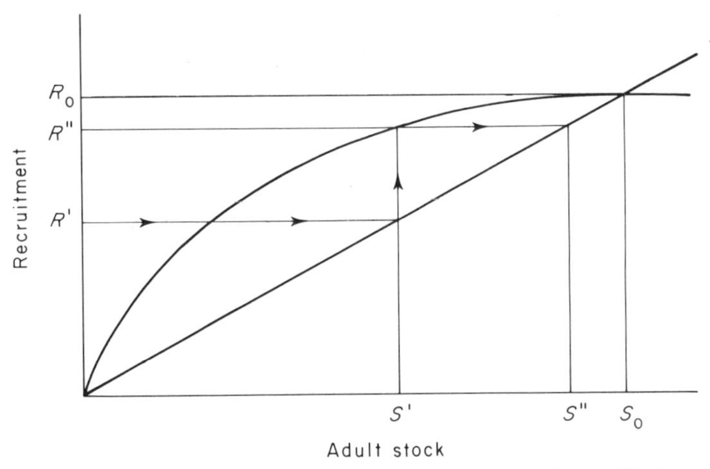

Figure 6.1 The pattern of the return of a stock towards the equilibrium position (R_0, S_0) following a disturbance, as determined by the relations between parent stock, recruitment, and resulting adult stock

We will assume at first that the environmental factors affecting recruitment are constant. Then two relations between stock and recruitment can be considered. First, the average strength of recruitment that will arise from a given stock—this will be one of the family of curves discussed in section 6.4. The other relation is the stock size that will arise from a given recruitment. For a certain pattern of fishing, and assuming constant mortality and growth rates, this will be proportional to the recruitment, i.e. the relation is a straight line. These relations are shown in Figure 6.1. Where they cross gives the equilibrium values of stock and recruitment, S_0 and R_0. If the stock is disturbed from this point—e.g. by abnormal weather giving rise to a very poor recruitment, the stock will return in one or two generations (and in the absence of other disturbances) close to this equilibrium position. For example, a small recruitment R' will give a stock S'; this will give a larger recruitment R'', a stock S'', and so on.

Suppose the rate of fishing is increased. The adult stock arising from a given recruitment will then be smaller, and thus (bearing in mind that the independent variable is plotted along the y-axis) the line will be steeper. The point where this steeper line cuts the stock–recruitment line gives the new equilibrium point—see Figure 6.2. If the increase in fishing is moderate $(R_1 S_1)$, there will be little difference in recruitment. A higher rate of fishing $(R_2 S_2)$ will give rise to an equilibrium position with a significantly reduced recruitment, while at a still higher rate, indicated by the broken line in Figure 6.2, there will be no equilibrium. Attempts to exert this amount of fishing will soon lead to the collapse of the stock and the disappearance of the fishery.

To apply this approach in practice, the first steps are to determine the two relations. The curve comes directly from the methods of section 6.4. The line can be determined from the methods of Chapter 5.

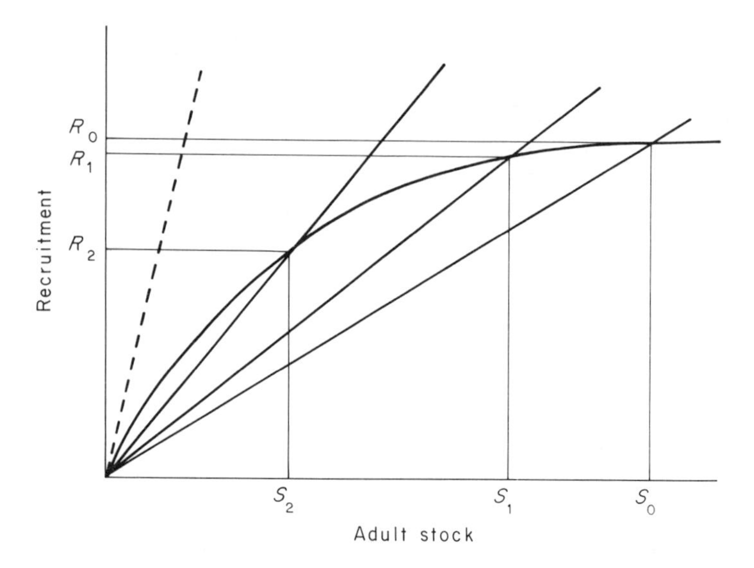

Figure 6.2 Equilibrium positions on the stock–recruitment curve for light (R_0, S_0), moderate (R_1, S_1) and heavy fishing (R_2, S_2), and the line corresponding to very heavy fishing with no equilibrium position

The size of the adult stock can be measured in a number of ways, but the most direct for the present purpose is the total egg production E. This can be given by

$$E = \sum_i k_i P_i q_i N_i \tag{6.12}$$

where k_i = average fecundity of fish of age i, P_i = proportion of females in the ith age-group, N_i = total number of fish age i alive at during the spawning season, and q_i = proportion of females age i that are mature.

Alternatively, since the fecundity per unit weight is often approximately constant i.e. $k_i = kw_i$, the total egg production can be given by

$$E = \sum_i kP_i q_i N_i w_i \tag{6.13}$$

Of the terms in equations (6.12) and (6.13), all except N_i will be in general independent of fishing. (The method can be readily extended to take account of a heavier fishing mortality on, say, females, causing P to vary with changing intensity of fishing.) Here, N_i will be given for a given recruitment R by the equation

$$N_i = R\exp\left[-\sum_{t_r}^{i} (F_t + M)\right] \tag{6.14}$$

in the general case where fishing mortality varies with age, or, if F is constant,

$$N_i = R\exp\left[-(F + M)(t_r - t_i)\right] \tag{6.15}$$

The procedure for calculating the equilibrium yield, allowing for the effect on recruitment, for any pattern of fishing, can now be set down:

(a) Plot the curve that is believed to best estimate the relation between stock and subsequent recruitment.
(b) From equations (6.14) or (6.15), calculate the line relating recruitment to resulting adult stock, for the given pattern of fishing.
(c) Note where the line and curve intersect, and the value of recruitment R_F at that point
(d) Calculate, by the methods of Chapter 5, the yield-per-recruit corresponding to the given pattern of fishing $[=(Y/R)_F]$
(e) Calculate the total yield Y_F as $Y_F = R_F(Y/R)_F$

This can be repeated for other patterns of fishing (values of F and age at recruitment), and hence curves obtained of total yield as a function of fishing effort, etc.

Valuable as these curves are as a much better guide to the likely effects of increased fishing than the yield-per-recruit curves of Chapter 5, too much meaning should not be attached to the results of using any one particular stock–recruitment curve. A greater value, in terms of understanding the likely behaviour of the stock and of providing long-term advice, comes from repeating the process, and especially steps (a) and (c) for different stock–recruit curves. The calculations need not be followed through completely for each possible stock–recruit curve. What is generally sufficient is to determine the point of intersection, and the effects on the equilibrium level of recruitment.

A useful way of looking at the available data is to plot the available pairs of observations of stock and recruitment (as should be done at an early stage in any study of stock–recruitment), and then, before fitting any curve, also draw in the recruitment–stock lines corresponding to important levels of fishing—particularly zero-fishing, the current pattern of fishing, and where relevant, the level of fishing likely to occur in a few years if no action is taken to control the level. Sketching possible stock–recruitment curves through the observed points, and noting when they cut the various lines can give immediate information on what might happen to the fishery, and the importance of examining the stock–recruit question in more detail. Figure 6.3 illustrates a typical example: curve (a) corresponds to constant recruitment over the ranges of stock likely to be expressed in practice; curve (b) suggests that up to the present fishing may have tended, on the average, to increase recruitment; while curve (c) suggests that the present level of fishing is sufficient to reduce the long-term change recruitment quite considerably, and that if effort is increased, and maintained at that level, there could be a collapse of recruitment. The implication of these results in relation to fisheries policy is that further expansion of fishing effort should be discouraged, at least until curve (c) is shown to be unlikely. The strength of the discouragement will depend on the outcome of the yield-per-recruit calculations; if these show that there would be little or no increase in yield-per-recruit from increased fishing, the discouragement should

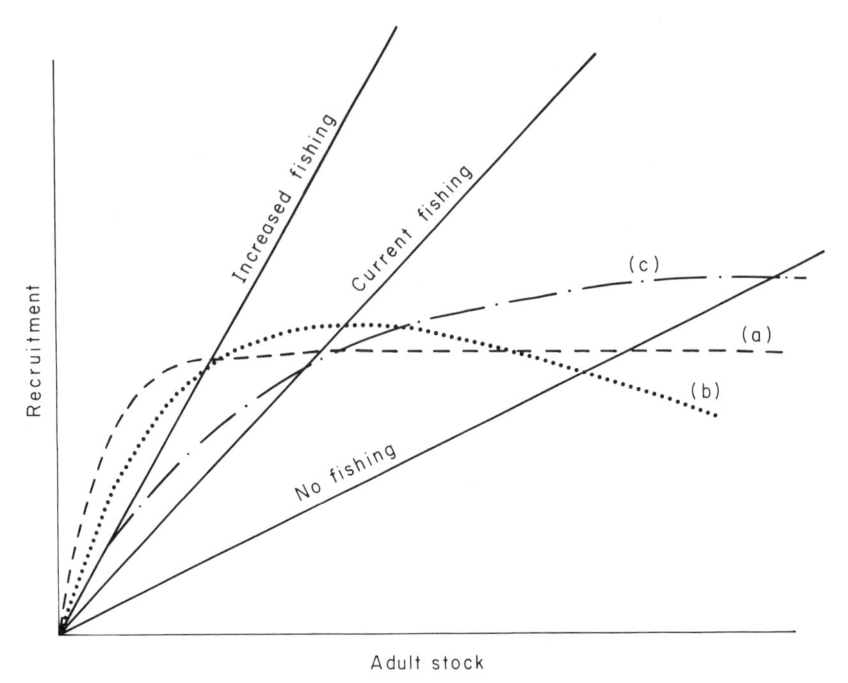

Figure 6.3 The effects of different stock–recruitment relations on the expected equilibrium position following increased fishing

be very strong. In relation to research policies and priorities, the results emphasize the need to look more carefully at the stock–recruitment relation, and especially to distinguish which of the curves is the more likely.

This graphical approach is also useful in looking at some of the effects of environmental variation. Instead of a single curve relating the parent stock to the average recruitment (or the recruitment under 'average' or 'normal' conditions), it is possible to consider a family of curves, each corresponding to a particular set of environmental conditions. In particular, there will be two limiting curves, corresponding to unusually good and unusually bad environmental conditions, which will follow the upper and lower bounds to the observed points. The confidence and precision with which these lines can be drawn will vary with the volume of observations available (essentially the number of years of data), but for the purposes of explaining the likely consequences of variability—and especially of determining whether or not the analysis based solely on examination of average conditions can give a dangerously incomplete picture of likely events—high precision is not needed. Equally, it is not essential to determine or define what is meant by 'good' or 'bad' environmental conditions—for example temperature at the time of spawning, food supply to the young fish, abundance of predators, etc.—but only to accept that in some years in the observed series of data, and therefore presumably some years in the future, recruitment falls well below that expected as an average, given the abundance of adults, and that in other years it is above.

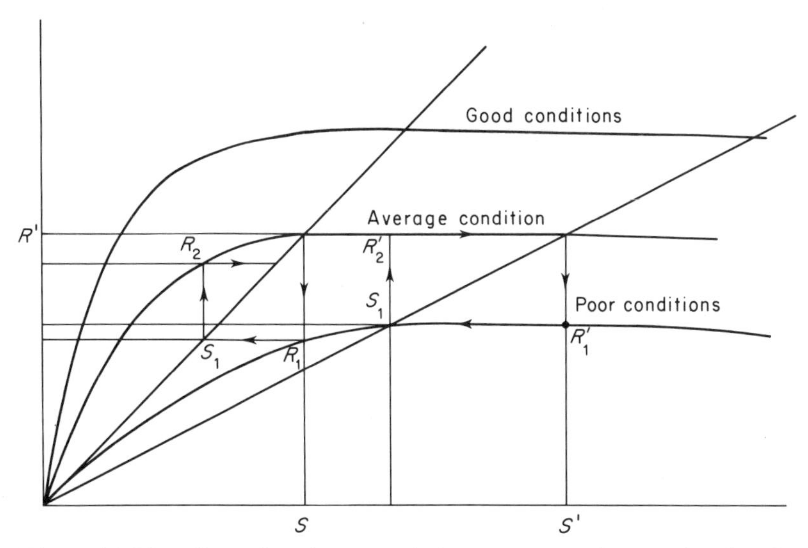

Figure 6.4 The effects of environmental changes on recruitment, showing the different returns towards equilibrium under light fishing (R'_1, S'_1, R'_2) and heavy fishing (R_1, S_1, R_2)

The type of picture that might emerge is illustrated in Figure 6.4. Apart from curves corresponding to 'poor', 'average', and 'good' conditions, lines corresponding to light and heavy fishing are also shown. Considering just the average curve, it can be seen that, if environmental conditions were always average, the stock would reach equilibrium positions (R', S, and $R'S'$ for the two levels of fishing), with recruitment in both cases approximately at the asymptotic level. One would then conclude that there was no need for fishermen or administrators to worry about the nature of the stock–recruitment relation. However, looking at what happens if there is one year of poor condition, followed by a return to average (it may be emphasized that 'average' should not be confused with 'normal', particularly in the sense that one should not expect the 'average' to occur every year or even most years), there is a big difference between heavy and light fishing. In the latter case the succession of values of recruitment and stock—R'_1, S'_1, R'_2—indicate a quick return to the equilibrium 'average' condition ($R'_2 \simeq R'$), whereas in the second case, the unfavourable conditions are sufficient to bring the stock on to the left-hand slope of the stock–recruit curve, as indicated by the succession $R_1 S_2 R_2$, and the recruitment does not immediately return to the 'average equilibrium' points, i.e. $R_2 < R'$.

The combination of yield-per-recruit and stock–recruitment studies can also be approached by computer simulation techniques (e.g. Walters, 1969; Fox, 1973). These do not attempt to arrive at an explicit expression of the equilibrium situation corresponding to any given pattern of fishing. Rather, they simulate the history of the fishery under each pattern of fishing over say 100 generations, by which time either an equilibrium position will have been arrived at or the system will show clear oscillations. These latter will be of considerable interest, and provide information on the likely character of the actual fish stock under the proposed pattern of exploitation which would be additional to any information about an equilibrium, or quasi-equilibrium position. Simulation models can also be used to compare observed events over some past period with what might have happened if one or other of the several factors involved had been changed (Sissenwine, 1977).

The logical steps involved in a simulation model are simple, and similar to those already discussed. Starting with a certain recruitment, its fate in the fishery is calculated using the given values of growth and mortality, including the fishing mortality at each age corresponding to the particular pattern of exploitation. This will determine, in addition to the yield of the fishery, the size of the spawning stock. The favoured stock–recruit relation (possibly with a random element to represent environmental effects) can then be used to determine the subsequent recruitment. In turn the fate of this year-class can be followed through the fishery, and the whole exercise repeated for as many generations as desired. This procedure can be readily changed to include longer-lived species, with repeated spawning. The calculations for each year then take account of all the survivors from the previous year, plus the recruits. The latter may be calculated from the adult stock sufficient years in the past to match the average time between birth and recruitment. As in any simulation model, further details can be added to this basic outline to take account of those specific aspects of the fishery being studied that are believed to be important.

The application of simulation models to the study of the yield under different patterns of exploitation and hypotheses concerning the stock–recruit relation are not discussed further here. The main reason is that such discussion, if it is to be useful, will require extensive exposition of the actual programs. Programs, which can be applied to particular fisheries with a greater or lower degree of modification, already exist, and their number is growing. It will in most cases be better to use one of these rather than write a new program; the choice will depend on the type of problem.

The differences in facilities between different countries and different institutions, and the rate of change in computers and in programs available, make it difficult to make any useful recommendations in a manual of this kind. Another reason is that in practice, given the uncertainties concerning the true stock–recruit relation in any particular fishery, it is doubtful whether a sophisticated computer model will give a more useful insight into how the fishery might behave than the graphical approach approved described earlier.

REFERENCES AND READING LIST

Allen, K. R. (1973) The influence of random fluctuations in the stock recruitment relationship on the economic returns from salmon fisheries. *Rapp. P.-V. Reun. CIEM*, **164**: 350–359.

Backiel, T. and E. D. Le Cren (1978) Some density relationships for fish population parameters. In *Ecology of Freshwater Fish Production*, S. D. Gerking (ed.). Oxford, Blackwell Scientific Publications, pp. 279–302.

Beverton, R. J. H. and S. J. Holt (1957) On the dynamics of exploited fish populations. *Fish. Invest. Minist. Agric. Fish. Food UK* (*Series* 2), **19**: 533 pp.
Clark, C. W. (1974) Possible effects of schooling on the dynamics of exploited fish populations. *J. Cons. CIEM*, **36** (1): 7–14.
Csirke, J. (1980) Recruitment in the Peruvian anchovy and its dependence on the adult population. In *The Assessment and Management of Pelagic Fish Stocks*, A. Saville (ed.). *Rapp. P-V. CIEM*, **177**: 307–313.
Cushing, D. H. (1971) The dependence of recruitment on parent stock of different groups of fishes. *J. Cons. CIEM*, **33** (3): 340–362.
Cushing, D. H. (1977) The problems of stock and recruitment. In *Fish Population Dynamics*, J. A. Gulland (ed.). London, Wiley–Interscience, pp. 116–133.
Cushing, D. H. and R. R. Dickson (1976) The biological responses in the sea to climatic changes. *Adv. Mar. Biol.*, **14**: 1–122.
Foerster, R. L. (1968) The sockeye salmon, *Oncorhynchus nerka*. *Bull. Fish. Res. Board Can.*, No. 162: 422 pp.
Fox, W. W. (1973) A general life history exploited population simulation with pandalid shrimp as an example. *Fish. Bull. NOAA/NMFS*, **71** (4): 1019–1028.
Gulland, J. A. (1975) The stability of fish stocks. *J. Cons. CIEM*, **37** (3): 199–204.
Le Cren, E. D. (1958) Observations on the growth of perch (*Perca fluviatilis* L.), over twenty-two years with special reference to the effects of temperature and changes in population density. *J. Anim. Ecol.*, **27**: 287–334.
Parrish, B. B. (ed.) (1973) Stock and recruitment. *Rapp. P.-Reun. CIEM*, **164**: 372 pp.
Parrish, R. H. and A. D. MacCall (1978) Climatic variation and exploitation in the Pacific mackerel fishery. *Fish. Bull. Calif. Dep. Fish Game*, No. 167: 110 pp.
Ricker, W. E. (1954) Stock and recruitment. *J. Fish. Res. Board Can.*, **11** (5): 559–623.
Rounsefell, G. A. (1958) Factors causing decline in sockeye salmon of Karluk River, Alaska. *Fish. Bull. USFWS*, **58**: 79–169.
Sahrhage, D. and G. Wagner (1978) On fluctuations in the haddock population of the North Sea. *Rapp. P.-V. Reun. CIEM*, **172**: 72–85.
Saila, S. B. and E. Lorda (in press) A critique and generalization of the Ricker stock–recruitment curves. *Can. J. Fish. Aquatic. Sci.*
Sissenwine, M. P. (1977) A compartmentalized simulation model of the southern New England yellowtail flounder (*Limanda ferruginea*) fishery. *Fish. Bull. NOAA/NMFS*, **75** (3): 465–482.
Walters, C. (1969) A generalized computer simulation model for fish population studies. *Trans. Am. Fish. Soc.*, **98** (3): 551–559.
Walters, C. J. and D. Ludwig (1981) Effects of measurement errors on the assessment of stock–recruitment relationships. *Can. J. Fish. Aquat. Sci.*, **38**: 704–710.

EXERCISES

Exercise 6.1

A certain species recruits at one year old, and may be harvested during the next four years of life. The average yield per 1000 recruits (in kg) during these years under three different fishing strategies is given in Table 6.1.

Table 6.1

Year of life	2nd	3rd	4th	5th
Light fishing	20	35	40	30
Moderate fishing	50	65	30	15
Heavy fishing	110	40	10	5

The numbers of recruit in 1965 and the 10 subsequent years were as follows (in millions): 3000; 2500; 500; 100; 7000; 1500; 3000; 2500; 1000; 6000.

Calculate the yield, under each strategy, in each year from 1968 to 1975. Which strategy gives the greatest total yield (or greatest average annual yield)? For which is the yield in the worst year of the period greatest? Which strategy might you recommend (a) if the fishery supplies a very elastic market and the fishermen have other sources of income, and (b) if the fishermen have few alternative sources of income.

Exercise 6.2

Table 6.2, taken from Raitt (1939), gives the lengths (in cm) of haddock in the central part of the North Sea at the end of each of their first four years of life. Table 6.2 also gives the strength of each year-class as the catch per hour's fishing by survey vessel during their second year of life.

Table 6.2

Year-class	Year of life: 1st	2nd	3rd	4th	Year-class strength
1925	—	22.7	27.9	31.7	146
1926	15.8	23.3	27.5	29.9	211
1927	14.9	23.4	27.8	31.1	40
1928	16.7	22.2	26.4	30.5	496
1929	16.1	22.9	28.0	31.1	79
1930	17.1	24.5	29.4	34.7	41
1931	17.2	24.5	29.7	33.4	221
1932	17.5	24.8	30.3	33.8	62
1933	17.0	26.0	31.6	35.6	77
1934	18.0	26.9	32.7	35.9	21

(i) Calculate the increment in length during the second to fourth years of life for each year from 1928 to 1934. (*Note*: The biological year is taken from April to March, so the 1925 year-class had its first birthday on 1 April 1926.)

(ii) Hence, calculate the parameters of the von Bertalanffy growth curve for each year: (a) making independent estimates for each year; (b) assuming that the value of K is 0.25, equal to that obtained by pooling all data.

(iii) If the density of haddock may be estimated from the catch per hour of British steam trawlers, which were as follows (in kg):

1928	35.6
1929	33.5
1930	38.8
1931	34.8
1932	33.8
1933	33.5
1934	24.8

examine the relation between the growth parameters and total density of commercial sizes of fish. Also examine the relation of growth to other indices of density, e.g. the strength of the year-class of fish in their second year of life.

Exercise 6.3

(i) The data for the plaice fishery in the North Sea given in exercise 5.2 shows that for the present fishing mortality ($F = 0.5$) increased fishing would decrease the yield-per-recruit, and a decrease of 50 per cent would increase the yield-per-recruit, provided natural mortality remains constant ($M = 0.1$). By calculating the yield-per-recruit for $F = 0.5$, $M = 0.1$, and the following sets of values:

(a) $F = 0.6$; $M = 0.1, 0.095, 0.09, 0.08, 0.07$.
(b) $F = 0.25$; $M = 0.1, 0.11, 0.12, 0.15, 0.20$.

find what density-dependent changes in natural mortality would be necessary: (1) for a 20 per cent increase in fishing to give an increase in Y/R; (2) for a 50 per cent decrease in fishing to give a decrease in Y/R.

(ii) Repeat (i) for some fish stock familiar to you. Also, calculate the changes in density (as estimated by $Y/R \cdot 1/F$). Do the changes in M that would be required to alter the conclusions based on constant M appear reasonable to you, in the light of what is known about the biology of the fish and the changes in density?

Exercise 6.4

Data of adult stock and subsequent recruitment from two fish stocks are given in Table 6.3 (figures are arbitrary units).

Table 6.3

	Species A		Species B	
Year	Adult stock	Recruitment	Adult stock	Recruitment
1	8.8	7.1	7.3	8.0
2	7.4	6.4	12.5	8.3
3	4.5	6.4	14.3	7.3
4	13.2	7.0	10.8	7.0
5	14.6	4.7	17.0	6.7
6	7.0	7.0	15.7	8.3
7	3.1	5.4	4.6	7.0
8	7.7	6.1	9.0	9.1
9	10.7	3.8	4.0	5.4
10	8.6	6.0	5.7	8.4
11	15.4	6.2	10.4	8.7
12	2.0	6.3	2.5	4.5

Examine the form of the stock–recruitment relation for the two species: (a) graphically, and (b) by fitting an appropriate curve.

For stock B, if a Ricker curve is fitted, what is the stock giving the maximum average recruitment, and what is the expected recruitment from that stock?

What advice would you give to managers, for each species, if the present stock was (a) 10 units, (b) 15 units, (c) 2 units?

How would your answers change if, for stock A, the stock and recruitment for two later years were 0.6; 4.5 and 0.4; 3.0 respectively?

Exercise 6.5

Table 6.4

Year	Haddock S	Haddock R	Mackerel S	Mackerel R	Shad S	Shad R	Salmon S	Salmon R
1921	39	21					150	449
1922	44	5					40	228
1923	76	281					70	199
1924	72	119					106	81
1925	55	96					162	161
1926	91	155					253	146
1927	104	31					87	162
1928	73	351	177	93			109	263
1929	62	66	247	200			90	159
1930	39	30	240	206			109	117
1931	77	183	349	321			87	258
1932	49	32	448	230			74	254
1933	57	70	634	85			97	219
1934	44	25	623	37			145	126
1935	27	209	576	47			88	125
1936	20	26	358	96			137	135
1937	17	19	225	69			126	133
1938			189	115			123	159
1939			159	96			71	183
1940			150	90	212	193	88	86
1941			132	159	284	188	93	57
1942			126	72	237	156	63	69
1943			185	61	229	135	92	150
1944		58	165	53	200	113	77	114
1945		186	124	19	123	83	66	126
1946	165	2	101	20	101	103	44	82
1947	62	22	57	131	75	129	48	77
1948	57	47	43	61	86	114	75	81
1949	55	39	92	10	75	97		
1950	41	81	96	4	65	80		
1951	33	388	65	3	89	85		
1952	31	120	32	62	120	128		
1953	25	125	28	125	74	163		
1954	29	154	58	50	77	166		
1955	64	133	105	82	59	169		
1956	77	10	100	21	138	157		
1957	68	5	85	24	151	123		
1958	96	291	39	88	149	119		
1959	74	32	54	59	154	133		
1960	35	29	80	127	149	129		
1961	32	177	97	87	135	125		
1962	23	2815	159	14	84	105		
1963	31	2	155	8	101	112		
1964	164	5	104	2	130	138		
1965	322	8	38	4	121	144		
1966	260	165	18	1	133	109		
1967	108	1659	4	1	110	111		

Table 6.4 (*contd.*)

	Haddock		Mackerel		Shad		Salmon	
Year	S	R	S	R	S	R	S	R
1968	18	319	3	4	115	128		
1969	93	102			160	168		
1970	442	143			160	227		

Table 6.4 gives data on adult stock (S) and subsequent recruitment (R) for four typical stocks:

(a) North Sea haddock—stock as egg production. Recruitment as catch per hour of age I fish by Scottish trawlers. Data from Sahrhage and Wagner (1978).
(b) Pacific mackerel, in the California current—stock as spawning biomass (millions of pounds). Recruitment as biomass of age I recruits (millions of pounds). Data from Parrish and MacCall (1978).
(c) American shad—Spawners (escapement) and recruits as thousands of fish. Data from Grecco (unpublished report), quoted by Saila and Lords (in press).
(d) Sockeye salmon in Karluk Lake, Alaska—spawners (upstream escapement) and recruitment as tons of thousands of fish. Data from Rounsefell (1958), quoted by Foerster (1968).

(1) Examine these data for general evidence of a relation between adult stock and subsequent recruitment: (a) graphically (a logarithmic scale would be useful for some stocks); (b) by constructing 2 × 2 or 3 × 3 tables, dividing stock and recruitment in halves (above or below the median) or thirds, and applying χ^2 or other test to these tables.
(2) Fit appropriate stock–recruitment curves to these data. What are the parameters of the Ricker curve? (Fit a linear regression of $\log S/R$ on S.)

Exercise 6.6

The haddock data of exercise 6.5 are highly scattered, and admits of different interpretations. Plot these data, and draw three curves:

(a) The Ricker curve obtained in example 6.5 by linear regression of $\log R/S$ on S.
(b) That curve which is consistent with the data, but which agrees best with the hypothesis that average recruitment is constant for all observed stocks.
(c) That curve which is consistent with the data, but which agrees best with the hypothesis that average recruitment decreases appreciably at the lower range of stock sizes.

[Curves (b) and (c) should be drawn by eye.]

If the line through the origin and the average recruitment and stock for the whole period can be taken as representing the relation between recruitment and resulting adult stock size under the average fishing conditions for the whole period, what would be the effect on average recruitment, in equilibrium conditions, of changes in the pattern of fishing that:

(i) increased the spawning stock from a given recruitment by 20 per cent;
(ii) decreased it by 20 per cent;
(iii) decreased it by 50 per cent?

Given that the results of exercise 5.3 describe the yield-per-recruit curve for haddock, what would be your advice to managers?

CHAPTER 7

Species interactions

7.1 WEAKNESSES OF SINGLE-SPECIES ASSESSMENTS

It is a common criticism by scientists interested in general and wide-ranging aspects of ecology of traditional stock assessment work that it is too often concerned only with what happens to a single species, and with the data (catches, catch per unit effort (c.p.u.e.), age composition, etc.) relating to that species, and that proper account is not taken of what is happening to other species with which the target species interacts (i.e. its food, its predator, and its competitors). One answer to such criticism is that one type of traditional single-species model—the production model—does implicitly take into account the interactions between species. For example, if one relates the observed changes in the c.p.u.e. of shrimp off Kuwait to the changes in effort, these observations should take account not merely of the direct impact of fishing but also any indirect effects due to changes in the food supply, etc. The likelihood that the observations do in fact take full account of the latter effects will be increased if adequate allowance is made for possible lags in the system (e.g. by relating c.p.u.e. in year x to average effort in years $x-2$ to x). The extent of lag that is used should depend on the typical life span of the species involved. The procedure is likely to be least satisfactory in dealing with possible interspecific effects when some of the non-target species are much longer lived than the target species, i.e. a single catch/effort or production-model analysis of a krill fishery with, say, a two-year lag period, would probably not take proper account of possible interaction with whales or seals. Conversely, analysis of a long-lived animal using a simple single-species model will probably take an adequate account of possible interactions with shorter-lived species.

This reply to ecological criticism is more than just a quibble, but it is not the proper reply. This is that any good stock assessment scientist would indeed like to take interspecific interactions into account. An understanding of how the target species fits, into the general ecosystem in which it lives should be one of the long-term objectives of any stock assessment work, but it is not the principal or most urgent one. Too much attention paid to the wider ecological questions can too easily mean inadequate attention being paid to determining the effect of fishing on the target stocks (whether of one or several species). The result can then be poor advice on the effects of changes in fishing strategy (development or management). In the past the history of analysis and advice based on single-species approaches to what have often been essentially single-species fisheries has been moderately satisfactory. Poor advice has generally been due to inadequate data, or poor analysis, rather than to any failure to consider explicitly the wider ecological context.

This comforting situation is changing with the increasing range of species being exploited by present-day fisheries. There is an increasing risk that analysis and advice based on a single-species approach will be incomplete in some important aspects, and will lead to misleading conclusions regarding the effects of different policies. Changes in the fishery on target species can have effects on other species, and to the extent that it is increasingly likely that the non-target species will themselves support significant fisheries, these effects may have important practical impacts. For example, increased shrimp trawling may reduce the catches of fisheries for larger species of fish because of the increased catches of juvenile fish. These effects can be positive as well as negative. For example, it has been suggested that heavy fishing of anchovy off California, even to the point of 'over-exploitation' in a single-species sense, could be desirable because it would lead to the recovery of the sardine stock.

Another reason for a single-species analysis giving rise to misleading conclusions is a change of interest by the fishermen from one species to another. The normal sets of data (e.g. series of statistics of c.p.u.e. of a given species) then cease to measure reliably the quantities that they are assumed to do in a single-species analysis. For example, statistics of c.p.u.e. of a given species will not provide a satisfactory index of the abundance of that species if it ceases to be a preferred target of the fishermen. Also, fisheries on other species can affect the stock of the target species, so that the conclusions from a single-species analysis may become somewhat irrelevant. For example, it is not particularly helpful to consider ways of managing a croaker or snapper fishery if the main factor controlling the success or otherwise of that fishery is the amount of juvenile croakers or snappers occurring as by-catch in a shrimp fishery in the same region.

Such various interspecific interactions can therefore make a radical difference to what should be chosen as the optimum fishing policy. It is therefore well recognized that stock assessment and advice to policy-makers should take them into account (see FAO, 1978 for a review of some of the problems). However, it is less clear what can be done in terms of practical analysis, as shown by the lack of clear conclusions in many theoretical examinations of the problem (e.g. Hobson and Lenarz, 1977; May *et al.*, 1979) and by the manner in which there have been major changes in such well-studied areas as the North Sea, which are not easily explained in terms of single-species models (see Hempel, 1978). The following sections are therefore less of a manual on how species interactions should be taken into account than a general discussion of ways in which the problems can be approached.

7.2 BY-CATCHES

The simplest form of interaction between fisheries on different species is purely technological. Fishermen interested primarily in one species may include other species incidentally in their catches. These by-catches may be an important part of the total catch of the incidentally caught species, and also (at least in terms of weight) in the catches of the directed fishery. The most important examples are the fisheries for tropical shrimp (mostly penaeids). In some of these fisheries, for example those in the Gulf of Mexico, shrimp may make up little more than 10 per cent of the catch, at least in some places and times, with the rest being fish of various species and sizes. These fish include small individuals of species, e.g. croakers, which might grow to a large and valuable size.

The occurrence of by-catches raise a large number of biological, technological, economic, and social questions. Inevitably the directed fishery (e.g. on shrimp) will have some adverse effects on any fishery on the by-catch species (e.g. croaker). If the size of fish in the by-catch is below the optimum size at first capture for the current overall rate of fishing, the occurrence of the by-catch will reduce the total catch of the by-catch species, as well as transferring some of this total from one group of fishermen (for whom that species may be of prime importance) to another, for whom it may be of only marginal interest. Since the two groups of fishermen are likely to use different gears and types of vessel, and often come from different ports or from different countries, the potential for conflict is obvious.

One line of attack in reducing this conflict has been technological, to modify the methods of fishing so as to reduce the amount of by-catch. A successful example of this has been reduction in the number of porpoises killed in the eastern Pacific purse-seine fishery for yellowfin tuna.

The trawl used for catching shrimp (mostly pandalids) in northern waters has been adapted to reduce the catch of small fish (FAO, 1973), but a practical adaptation of the same approach for tropical (penaeid) shrimp has not so far been found. A different technical approach to the shrimp by-catch problem has been to develop methods for using those fish which are in many cases (especially in the case of the larger shrimp trawlers) discarded dead at sea.

The by-catch is not always of negligible interest to the directed fishery. The value of the minor species can make all the difference between a profitable and a non-profitable trip. Then the problem may be less one of reducing or eliminating the by-catch than of maintaining it, for example, in the sub-arctic trawl fisheries of the Atlantic the main target species (cod) grows to a bigger size than several of the other species (e.g. haddock) taken. The mesh size used in these fisheries were, on the whole, much smaller than the optimum one for cod. The introduction of larger meshes for the cod fisheries met difficulties because these would cause losses of haddock, and other species, which might more than counterbalance the gain in cod catches.

These difficulties, however, concern a political and economic choice, or are matters of technology. The stock assessment work is straightforward, once it is remembered that when studying the population dynamics of a stock of fish *all* catches from that stock have to be taken into account. Equally, when assessing the effects of proposed changes in a fishery (increasing the amount of fishing, altering the mesh size used), the effects on *all* species taken in that fishery should be considered.

It can be difficult to collect detailed data in respect of all the fisheries and species, and in practice 'taking into account' or 'considering' may involve little more than a rough approximation to the catches involved, and their characteristics (size composition). This is acceptable, especially if there is enough information to show that the total quantities involved are small. What is not acceptable is to ignore them without examining the information. In summary, the major step in respect of influence of possible by-catches when assessing a given fishery, for example the Scottish North Sea trawl fishery for haddock, is to answer these two questions:

(1) What other fisheries catch fish from the same stock (e.g. Scottish seine-net fishery for haddock, the Danish fish-meal fishery)?
(2) What other species are taken in the same fishery (e.g. whiting, cod)?

Once these questions are answered, and the data collected, then the assessments for each species can proceed in accordance with the methods of the previous chapters. In particular, taking F_i' and F_i'' to denote the mortalities on fish age i in the directed fishery, and as by-catch in the other fisheries, equations (5.15) or (5.16) can be used to calculate the yield-per-recruit.

What can be considered as a special case of the by-catch problem occurs when a fishery exploits several stocks on an approximately equal footing; for example, a salmon fishery which at the mouth of the river can catch fish from a number of different parent streams. Given methods of distinguishing the origin of the fish in the catch, the separate assessment of each stock presents no problem of principle, but if the fishery can be selectively diverted to one stock or the other (which may not be easy), the management of such a fishery, and the determination of the optimum exploitation strategy can be complicated (see Ricker, 1958; Paulik *et al.*, 1967; Hilborn, 1976).

7.3 THE EXTENSION OF PRODUCTION MODELS

To take account of the biological interactions between different species, the simplest method is to extend the production models described in Chapter 3.

One method of doing this is to consider that the equation of population growth, e.g. equation (3.5)

$$\frac{dB}{dt} = f(B)$$

used to describe the dynamics of a single stock, for example haddock on Georges Bank, could be equally valid as a description of the dynamics of a group of species, taken together, for example all the demersal fish on Georges Bank. If this approach is valid, then the methods of Chapter 3 can be used directly, e.g. the linear regression of C/F on f of equation (3.9), where c/f is the c.p.u.e. or density, of all species, and f is the effort on all species.

The immediate attraction of this approach is its simplicity, and apparently small demands on data. In fact the data supply may not be simple. If various groups of fishermen have preferences for one or other species, then it is difficult to extract, from the normal commercial statistics on catch and effort, estimates of c.p.u.e. that apply to the stocks as a whole. The alternative is to derive an index of abundance independently of the commercial fisheries, e.g. from surveys by research vessels. Such surveys now exist, for a respectable number of years, in several areas (Georges Bank, Gulf of Thailand).

The index of abundance U, can be taken directly as the mean c.p.u.e. in the surveys, or as the mean density based on the sampling strategy (e.g. stratification) used in designing the survey. These are subject to some degree of bias, because whatever sampling gear is used it will be somewhat selective, catching some species more effectively than others. If changes in overall density are accompanied by changes in species composition, the c.p.u.e. or density as estimated from research catches will not accurately reflect changes in total abundance. If the selectivity of the survey gear is known, even approximately, a better index of abundance, $\bar{U}$, can be obtained as

$$\bar{U} = \frac{\sum 1/q_i \cdot U_i}{\sum 1/q_i} \tag{7.1}$$

where U_i = the c.p.u.e. for the ith species, and q_i = the catchability coefficient for that species.

In either case, equation (3.9) can now be used in the form

$$\bar{U} = a - b \cdot \frac{C}{U} \tag{7.2}$$

where a, b = constants, and C = the total catch.

The theoretical objections to this procedure are less serious than might be expected from its simplicity. The single-species production model, as fitted to observed data, measures the net effect on the stock of changes in the amount of fishing, taking account of the interaction with the whole environmental system or at least those that act quickly. Similarly, the observed relation between the index of abundance of all species, i.e. $\Sigma U_i, i = 1, 2 \ldots n$, and the total effort, as defined above, will encompass the combined effects of the various interspecific interactions. This is, however, only true provided the interactions act sufficiently quickly for the observed data to be representative of equilibrium conditions, and as long as no irreversible changes in species composition occur. It also holds only so long as the relative fishing intensity on different species remains the same, i.e. the fishermen do not change their preferences or effective selectivities in respect of different species. For this reason this approach may be more suited to the typical tropical fisheries on a large number of short-lived species, where selection between species is difficult, than for temperate water where fishermen may make a deliberate selection between a handful of species.

In practice it has been found that the fit to observed catch and effort data is often better in terms of the combined total of all species (or all demersal species) than it is in terms of species taken individually. This may be because indeed the approach is better, in that total biomass reacts in a simple way to fishing, and the various interspecific reactions are taken into account, or at least the available effort data better reflects the general fishing mortality on all species than the mortality on individual species. Alternatively, the improvement may only be an artefact, arising for example, from the averaging process. For the present it can be said that the approach is useful, but (as indeed is the case for any assessment technique) should be as far as possible checked by alternative approaches, such as for example, by following changes in the species composition that will undoubtedly occur as a function of effort.

Another method of extending the production model to several species is to consider the equation describing the dynamics of each species. If each of these species are considered in isolation we have from equation (3.5)

$$\frac{dB_1}{dt} = f_1(B_1) \tag{7.3}$$

$$\frac{dB_2}{dt} = f_2(B_2) \tag{7.4}$$

Taking the simplest forms for $f(B)$, and rewriting equation (3.6), we have

$$\frac{1}{B_1}\frac{dB_1}{dt} = a_1 - r_1 B_1 \tag{7.5}$$

and

$$\frac{1}{B_2}\frac{dB_2}{dt} = a - r_2 B_2 \tag{7.6}$$

The simplest expression to take account of interactions between the two populations is to add a linear term to the right-hand side of these equations, give the well-known Lotka–Volterra equations (Volterra, 1926):

$$\frac{1}{B_1}\frac{dB_1}{dt} = a_1 - r_1 B_1 - C_{12} B_2 \tag{7.7}$$

$$\frac{1}{B_2}\frac{dB_2}{dt} = a_2 - r_2 B_2 - C_{21} B_1 \tag{7.8}$$

If C_{12}, C_{21} are both positive, this pair of equations can be taken to represent competition between species. If C_{12} is positive, and C_{21} negative, i.e. the second species benefits from the presence of the first, this can be taken to represent predation of the second species on the first.

Theoretically these equations can be extended to cover any number of species, but beyond two species the practical problems of fitting them to actual observations increase enormously. For n species these are $n(n+1)$ parameters (n values of a and r, and $n(n-1)$ values of C). In tropical fisheries, with 50 or 100 species, this means estimating several thousand parameters, a number that is likely to far exceed the number of degrees of freedom.

In practice it is doubtful whether this approach can be carefully applied to more than two or three species, or two or three groups obtained by lumping species together. For two species it is convenient to use expressions in terms of the quantities for which estimates are most likely to be available—the fishing effort (f_1,f_2) on each species, and the indices of abundance (U_1 U_2), available from survey data or the c.p.u.e. in a fishery directed at the species concerned. It may be noted here that direct estimates of the total fishing effort (taking account of all types of vessel and gear) is usually difficult. The effort, f_i, on each species will be usually estimated as $f_i = C_i/U_i$. These estimates can be incorporated in the linear expressions comparable to equation (3.3).

$$a_1 - b_1U_1 - c_{12}U_2 - f_1 = 0 \tag{7.9}$$

$$a_2 - b_2U_2 - c_{21}U_1 - f_2 = 0 \tag{7.10}$$

Multiple regression techniques may then be used to provide estimates of the six parameters $(a_1, a_2, b_1, b_2, c_{12}, c_{21})$.

7.4 THE EXTENSIONS OF ANALYTIC MODELS

The analytic models also offer immediate opportunities for theoretical extensions to take account of interspecific interaction. Each of the parameters—R, M, K, W_∞ (or other growth parameters describing growth)—which are treated as constant in the simple methods of Chapters 4 and 5, can be considered as functions of the abundance of other species.

Some of the functions are obvious. The natural mortality is clearly related to the abundance of predators, and it is probably reasonable as a first approximation to consider mortality due to predation as being proportional to the abundance of predators. That is, we can write

$$M = M' + aB_p \tag{7.11}$$

where M' = natural mortality due to causes other than predator, and B_p = biomass of predators.

This leaves open the question of the values of M' and a. An idea of the value of a—which perhaps should not be termed an estimate—may be obtained from information on the food consumption—numbers per unit time per individual predator—of the predator population.

Similarly, the growth of the predators is related to the abundance of prey. This relation might be expressed empirically; for example, by a linear relation between W_∞ and prey abundance. A more satisfactory method is to note that of the food consumed only a part is digested. Of this part, some is used for maintenance, some (in adults) for reproduction, and only the rest for growth. Most of the parameters involved—maintenance requirements at different activity levels, efficiency of food conversion, etc.—can be fairly readily obtained from suitable experiments or otherwise. It is beyond the scope of the present manual to set out the resulting equations describing growth as a function of food input, and related aspects of the interactions between population of prey and predators. This can be found in numerous publications from the later chapters of Beverton and Holt (1957) onwards (see especially Andersen and Ursin, 1977); it should simply be noted here that this type of model exists, and that it provides the description of an interaction between different species which is among the most likely to bear a fair similarity to what occurs in any specific situation.

Predation of one fish species on another when both are of a fishable size is, however, not the only interspecific effect; nor is it probably the most important effect, either in terms of the magnitude of the changes, or the degree to which the changes can affect—or even make nonsense of—assessments based on simple assessments of each species individually. Just as the most important intraspecific effect is that on recruitment (see section 6.4), so the most important interspecific effect is also on recruitment. Also, here again, it is easier to build models to describe certain hypothetical events than to be sure that the actual events in the sea will follow, even approximately, the predictions of the model.

Two cases can be distinguished—competition between two species during part or whole of the pre-recruit phase, and predation of adults of one species on eggs, larvae, or other very young stages of a second species. The first type is less likely to cause surprises. Often the species that are competing at the larval, or other very early stages, are the same that are competing—in a manner that will be more obvious to the fishery scientist—at a later stage. For example, it is generally accepted that there is competition between various shoaling pelagic species (sardines, anchovies, mackerels, etc.) and that the decline of one of these species, due perhaps to overfishing, may be accompanied by the increase of another, even to the extent that the total production of, and catches from, the entire group of species remains approximately constant (e.g. Nagasaki, 1973). There is less general acceptance of the mechanisms of competition, but it may well be that the more important effects occur during the pre-recruit stages (see exercise 7.3).

The second type can give bigger surprises. Big fish like cod are well recognized as predators on small fish like herring, but when the cod are small enough the roles can be reversed. Few fish that eat zooplankton seem to be very discriminating about what types of zooplankton they eat. Each species has preferences, but a mackerel or a herring would be quite happy to include cod eggs or small larvae in their diet. Since the pelagic fish are usually abundant relative to the larger species, it does not take a large consumption of cod larvae by the individual mackerel to have a potentially large influence on the resulting year-class of cod. Interactions of this type have been put forward to explain the fact that during the late 1960s and 1970s, the year-classes of many commercially important North Sea demersal species, including cod, were well above average at a time when the stocks of mackerel, and later herring, had been seriously reduced by fishing.

If indeed reduced mackerel and herring stocks result in more cod, haddock, and plaice, as the discussion above suggests as possible, the practical implications are great. The demersal fish fetch, in general, a higher price than the pelagic species, especially if, as has been the case in the past, much of the latter are used for reduction to meal and oil. It would appear undesirable to apply restrictive management measures to rebuild the stocks of the pelagic species fully to their original levels if one result would be a fall in the catches of the demersal species. On the other hand, if the interactions were not as strong as suggested, it would be unfortunate to fail to rebuild the pelagic stocks in the false expectation of continued benefits from the other species.

At this point the weaknesses of present techniques of multi-species model-building and assessment become clear. A small change in one of the parameters, for example in the quantity of larval cod eaten by an individual mackerel, can make a big difference to the predictions of the changes in cod year-class strength caused by changes in the abundance of mackerel. The precision with which many of these important parameters can be estimated, either from empirical fits to the observed data, or from more direct observations such as the number of cod larvae found in mackerel stomachs, is unlikely to be high. Even if the basic model is correct, the confidence limits on any predictions (e.g. of the change in cod catches from a particular action concerning the mackerel fishery) are likely to be so wide because of the difficulties in estimating parameters that the practical value of the predictions will be small.

The growing number of detailed analytic multi-species models (e.g. Lett and Kohler, 1976; Andersen and Ursin, 1977; Laevastu and Favorite, 1977; Laevastu and Larkins, 1981) should therefore be considered as providing information on how the fisheries might behave, rather than how they *will* behave. This does not imply that they are useless. On the contrary, the qualitative results of most of these models show quite clearly that, however uncertain the precise behaviour of the system may be, it is reasonably certain that the fisheries will not react precisely as predicted by the single-species assessments. Further, the directions of departure from the single-species assessment are likely to be correctly predicted by the multi-species models—though it does not follow that even the most complex model will predict all the departures from single-species studies. For example, the possibility of the outburst of demersal species, including cod, in the North Sea in the 1970s being due to reductions of mackerel and herring is now accepted as a not unreasonable, if still unproven hypothesis, but the possibility does not appear to have been suggested before the striking rise in the abundance of most gadoid species became obvious, and some hypothesis to explain the outburst became necessary.

Since the predictions of these multi-species models—at least with the limited data available now, or in the foreseeable future—are essentially qualitative rather than quantitative, the important part of these models, so far as making assessments and providing advice is concerned, is their general form and the nature of the interrelationships that are taken into account. This is comforting, since the work of constructing a model, if it is to include writing and testing the necessary computer programs and estimating all the necessary parameters, is extremely lengthy and probably outside the capacity of most national fishery research institutions. On the other hand, most institutions do have the capacity to examine the information about their fisheries and, using the ideas of the available multi-species models, to determine in what direction the actual behaviour of the stock is likely to depart from the simple single-species predictions, and in qualitative terms ('slightly', 'moderately', 'strongly', etc.) the likely extent of this departure.

What is essential and cannot be avoided by using simulation models, or other techniques, is careful thought of how interspecific reaction could affect the fisheries being studied. This thought is helped by following a check-list such as that set out below. Such a check-list will reduce—but not eliminate—the possibility of important interaction being overlooked. This should cover the whole area in which each fishery of interest operates:

(i) What species are being fished? Which are the ones that are currently of major importance? ($S_1, \ldots S_s$)
(ii) What other species occur in the area and are of potential ecological importance? (This is not an exactly defined term, but can be taken to include all species that are numerous or large, or are likely to interact significantly with the species listed in (i).) Drawing up this check-list is not simple. A preliminary list can be readily obtained from commercial catches, or from survey results, but to this must be added species that may not be adequately sampled by the gear used. Information on the presence of such species can come by sampling with other gears, examination of the stomach content of predators, occurrence of eggs, or larvae in plankton hauls, etc.
(iii) What is know of the main ecological characteristics (distribution, food requirements, etc.) of each species?

Steps (i), (ii), and (iii) represent the collection of the basic material from which intelligent hypotheses can be made about interspecific reactions. Unless this information is reasonably complete, the resulting hypotheses are likely to be wrong, for example in failing to predict the effects of a competing but unexploited species which, because not caught commercially, was omitted from step (ii).

The next steps should be carried out for each commercial species separately, or at least, if a large number of species are taken (as in the case of many tropical fisheries), groups of species that might be expected to react in similar ways. For each species (or species group):

(iv) What species may compete with it? ($C_1, \ldots C_c$)
(v) What species eats it? ($P_1, \ldots P_p$)
(vi) What species does it eat? ($F_1, \ldots F_f$)
(vii) What species could affect its recruitment? e.g. through predation on larvae? ($R_1, \ldots R_r$)
(viii) What species might have its recruitment affected by S_1? ($R'_1, \ldots R'_r$)

These lists, in (iv) to (viii), should include all the interactions that need to be taken account in modifying the assessments of stock S_1 based on the single-species approach. The next stage is to put some kind of weights to the interactions. These weights should be both biological and practical.

On the biological side estimates (or reasoned guesses) should be made of the possible magnitude (or rather range of magnitude) of the effects, and the likelihood that they will occur. For example, the reduction in the stock of S_1

will, with a fair degree of certainty, decrease the natural mortality of its prey species, F_i, by the amount equal to reduction in the quantity eaten. This has readily calculable results, except for the possibility, which should not be ignored, that predators may find it easier to catch the sick or unfit members of the prey population. The reduction may therefore have a smaller long-term effect than might be expected. [In terrestial ecology the roles of predators in maintaining the health of a prey population is widely recognized.]

If several very similar species seem to be competing then the reduction in one might be expected to result in a roughly equivalent increase in the others. The more potential competitors there are, the more likely it is that such an equivalent substitution will occur. The Japanese pelagic fishery is based on half a dozen major species, and declines in one species have generally been balanced by a recovery by another, while no fish species seem to have benefitted from the decline in Atlantic–Scandinavian herring.

The immediate practical concern is naturally with those of the interacting species that are currently being fished, but attention must also be given to species that could support fisheries. If a species not currently of major commercial interest may increase as preferred species decrease, it is important to know the potential economic value of the replacement species. The relative values of different species are variable, as are the costs of catching them (a similar species that forms much smaller and more dispersed schools may be a highly unsatisfactory substitute for a pelagic fish that forms large schools, and is thus caught cheaply). Another point of practical interest is the identification of the groups of fishermen concerned with the various species. The concern of governments with the impact on non-target species of some development depends on who is interested in those species. It will be greater if the latter species form a major component of the catches of a group of small-scale fishermen than if they are mainly caught in an industrial-scale fishery with alternative supplies. Similarly, there will be less concern about possible changes in a community of pelagic species (anchovies, sardines, etc.) if the major fishery is for reduction to meal and oil (in which the exact species is not important) than if most of the catches are used for canning.

Besides considering the effects of possible changes in the fishery for species S_i on non-target species, it is also important to determine what changes in fisheries for other species are occurring, or are being considered, that might affect species S_i. For example, are there plans for the development of a big fish-meal fishery on species F_i, which might mean a reduction in the food supply of S_i, and hence in its growth rate?

All these considerations can then be summarized in two sets of statements of the following form, which can be used to qualify the single-species assessments and the advice arising from them:

(a) "The changes in S_i, arising from development or management decisions, will have the following effects (to be specified) on each of the following species (to be specified), and the fisheries on them." The listing will show whether the effects are positive or negative, roughly of what magnitude, and the confidence with which the prediction is made.

(b) "Changes in the fisheries on other species (to be listed) is expected to have the following effects (to be specified) on species S_i."

7.5 CONSIDERATION OF THE GENERAL ECOSYSTEM

Fish population dynamics, and fish stock assessment, can and should be considered as one branch of general quantitative ecology. (For a review article that relates fish studies to other quantitative studies, see Conway, 1977.) It is therefore perhaps surprising that there has not been better interchange between fishery scientists and ecology generally (see e.g. Werner, 1979), but there are reasons for this. One is that the two have developed at rather different speeds, governed to a large extent by the ability to collect data. Fish population dynamics was given a great boost by the large volume of data (particularly on catches and fishing effort, but also on the length and age composition of the catches) arising from the commercial fisheries. These data gave the incentive and opportunity for the notable advances in fish population dynamics that occured just before, and, more particularly in the period immediately following, the Second World War. Towards the end of that period, around 1955, there was a better quantitative understanding of the dynamics of exploited fish population than, probably, of any other group of animals. This situation is now largely reversed. The costs and practical difficulties of collecting data from marine fish stocks mean that fish dynamics still rely mainly on the same types of data from commercial fisheries, though routine surveys, carried out over periods of years, such as those in the Gulf of Thailand and on Georges Bank are becoming an increasingly useful source of data when the value of the resources being studied justifies the cost. Terrestrial ecologists on the other hand have found it easier to build up a more varied data base. Another new factor is the ability of computers to handle large amounts of data, and, in the present context more important, to examine implications of different theoretical models.

An important difference between fish stock assessment and general quantitative ecology has been the types of questions for which answers have been sought. The fishery questions have usually been of immediate tactical interest—How much catch should be taken next year to maintain a given stock in a healthy condition? The general ecologists have addressed more strategic questions—How does a population maintain itself at a given level over a long period? Some divergence in approach has been inevitable, and the fishery scientists have some excuse for not taking a great interest in other branches of quantitative ecology.

This situation is changing. On the one hand, the greater intensity of fishing, and the wider range of species being harvested, can cause sufficient changes in the ecosystem and the balance between species for the answers to the short-term tactical questions to provide a dangerously incomplete picture of the changes in the fishery over the medium to long term. On the other hand, the understanding achieved of the general behaviour of the ecosystem has advanced sufficiently to provides useful insight into how fisheries might behave. Concepts such as those of stability (e.g. May, 1973) are obviously relevant to understanding the sudden collapses such as that of the Peruvian anchoveta fishery in 1972. Though the ecologists may be still a long way from being able to predict at what point, for example with what combination of fishery and unfavourable environmental factors, a collapse is likely or inevitable, the knowledge that such a collapse is a typical feature of many populations under stress should be a great incentive to fishery scientists to look beyond the simple extrapolation of past experience. Similarly, the ideas of community structure can give insight into the likely behaviour of groups of species (e.g. sardines, anchovies, and mackerel) when all, or only some of them are subjected to heavy exploitation.

Most of these ecological concepts provoke thought, and suggest the more fruitful lines of thought—and well-directed thought, rather than the unthinking application of formula, is the vital ingredient of useful stock assessment—but one line—the studies of the flow of energy through the system from sunlight to fish harvest—does offer some potential for immediate and quantitative results.

If the biological production in the area can be neatly divided into plants, herbivores, and first, second, and third-stage carnivores, the production at each stage can be estimated in succession as a function of the plant production P_p as follows:

$$\text{Herbivores production} = P_H = {}_pq_H P_p$$

$$\text{Production of first-stage carnivores} = P_1 = {}_Hq_1 P_H = {}_pq_{HH}q_1 P_H$$

$$\text{Production of second-stage carnivores} = P_2 = {}_1q_2 P_1 = {}_pq_{HH}q_{11}q_2 P_H$$

where ${}_pq_H$, etc. are the efficiencies of ecological transfer from one stage to the next. If the stage at which commercial harvesting takes place—usually as first stage (e.g. herring, sardines), or second stage (e.g. cod) carnivores, and the proportion of the total production at that stage which can be harvested are determined, then the potential fish harvest can be determined.

This method has been used to estimate the total world fish potential by Schaefer (1965), and, for different ecological regions separately, by Ryther (1969). The problem in applying it is that the estimates obtained are highly sensitive to the stage at which harvesting is assumed to take place, and the values used for ecological efficiency. The greatest practical use is therefore to give a rough idea of the potential from an area, and in particular to determine whether current catches are a significant proportion of this potential.

What can be considered as a special case of this approach of relating potential fish yield to the basic biological productivity has been developed for groups of bodies of water of similar characteristics, especially lakes. Within a given group, e.g. the lakes of the south-eastern United States, the characteristics—species composition etc.—will be similar, so that the stage at which the fish harvest is taken, and the ecological efficiency at the preceding stages will be similar.

Therefore, within a group, the ratio of potential fish harvest and basic production can be expected not to vary much. Further, this production itself may be closely related to simple characteristics of the water body.

Examination of available data for several sets of inland water bodies has shown this to be a fruitful approach. The fish harvest per unit area of lakes in a given region can be determined quite accurately from a knowledge of the mean depth of the lake, $\bar{d}$, and the nutrient content of the water, as given by its conductivity, k. That is, the potential yield from a given lake will be given by

$$Y = cA\,\mathrm{MEI}$$

where A = area of the lake, MEI = morpho-edaphic index = $k/\bar{d}$, and c = constant, which varies with latitude being higher in tropical Africa, than in the lakes of northern Canada.

This does not remove the need to carry out any assessments. The potential of some members of the group must be determined independently in order to determine the value of the constant c. In practice, detailed assessment of each of the selected water bodies may not be essential. To a useful first order of approximation it may be sufficient to note that some bodies are fished very intensely, and that the yield from them is likely to be close to the potential yield, at least for the given pattern of fishing (types of gear, species preference, etc.).

This comparative approach is particularly valuable for fisheries in small scattered water bodies for which a detailed assessment of each body individually would be impracticable. Equally, it is often difficult to apply very detailed measures to manage or develop the fisheries in these waters, so that the precision required is not large. Though its main application has been to lakes (Ryder, 1965; Henderson and Welcomme, 1974) and rivers (Welcomme, 1976) it has also been found useful for reef fisheries (Munro, 1974–1976) and shrimps (Turner, 1977).

This comparative approach suggests a modification to the production model. Considering the c.p.u.e. as a function of fishing intensity (i.e. fishing effort per unit area), we have

$$U_i = a - b\frac{1}{A_i}f_i \tag{7.12}$$

where A_i = area of the water body concerned.

As normally applied, this equation refers to observations made of a given fishery at various periods, U_i being the c.p.u.e. in the ith period. However, provided the same relation between c.p.u.e. and fishing intensity hold good for different water bodies, then the equation can be applied to observations taken at the same time for different water bodies, the subscript i then referring to the ith area. The mathematical procedures for fitting are then identical.

A further extension to this method can be made to deal with the operations of different types of gear, each with its own selectivity in respect of species and sizes of fish. Adapting the method of Marten (1979), the c.p.u.e. of each gear can be expressed as a function of the efforts of all types of gear operating in the area. Assuming the simple linear form of relation this gives, for the ith area

$$U_{j,i} = a_j - \sum_k b_{jk} \frac{1}{A_i} f_{k,i} \tag{7.13}$$

where U_{ji} = the c.p.u.e. of the jth gear, in the ith area, f_{ki} = the effort of the kth gear in the ith area, and a_j, b_{jk} = constants, the same for all areas.

For only one gear, this reduces to equation (7.12). Given sufficient sets of values of U and f for different areas, estimates of a and b can be obtained by multiple regression. If there are m different gears, there will be m separate sets of multiple regressions, one for each gear. For each set the dependent variable, the c.p.u.e. of the chosen gear, as observed at, say, n different areas, will be considered as a function of the m independent variables $f_{k,i}/A_i$, the fishing intensity of each gear in each area. In principle this method enables complex interactions between different gears and different species to be taken into account. However, much depends on the similarity between different areas; the validity of the method probably requires a closer similarity than less detailed approaches, such as the use of the morpho-edaphic index. It may be noted that the parameters b_{jk} will normally be positive, i.e. the c.p.u.e. of any one gear will generally be reduced by increased fishing intensity by any other gear, especially if they both operate on the same species, but this is not necessarily always the case. For example, catches of small species by shore seines may be increased by the reduction of predators due to increased fishing by lines.

REFERENCES AND READING LIST

Andersen, K. P. and E. Ursin (1977) A multi-species extension to the Beverton and Holt theory of fishing, with accounts of phosphorus circulation and primary production. *Medd. Dan. Fisk.-Havunders.* (*New. Ser.*), **7**: 319–435.

Beverton, R. J. H. and S. J. Holt (1957). On the dynamics of unexploited fish population *Fish. Invest. Minist. Agric. Fish. Food UK* (*Series* 2), No. 19: 533 pp.

Conway, G. R. (1977) Mathematical models in applied ecology. *Nature, Lond.*, **269**: 291–297.

FAO (1973) Report of the Expert Consultation on selective shrimp trawls. Ijmuiden. The Netherlands, 12–14 June 1973. *FAO Fish. Rep.*, No. 139: 71 pp.

FAO (1978) Some scientific problems of multispecies fisheries. Report of the Expert Consultation on management of multispecies fisheries. Rome, Italy, 20–23 September 1977. *FAO Fish. Tech. Pap.*, No. 181: 42 pp.

Hackney, P. A. and C. K. Minns (1974) A computer model of biomass dynamics and food competition with implications for its use in fishery management. *Trans. Am. Fish. Soc.*, **103** (2): 215–225.

Hempel, G. (ed.) (1978) North Sea fish stocks—recent changes and their causes *Rapp. P.-V. Reun. CIEM*, **172**: 449 pp.

Henderson, H. F. and R. L. Welcomme (1974) The relationship of yield, morpho-edaphic index and number of fishermen in African inland fisheries. *CIFA Occas. Pap.*, No. 1: 19 pp.

Hilborn, R. (1976) Optimal exploitation of multiple stocks by a common fishery: a new methodology. *J. Fish. Res. Board Can.*, **33** (1): 1–5.

Hobson, L. S. and W. H. Lenarz (1977) Report of a colloquium on the multi-species fisheries problem, June 1976. *Mar. Fish. Rev.*, **39** (9): 8–13.

Jenkins, R. M. (1968) The influence of some environmental factors on standing stock and harvest of fishes in U.S. reservoirs. In *Proceedings of the Reservoir Fishery Symposium*, 5–7 April 1967. Athens, Georgia, Southern Division, American Fisheries Society, pp. 218–232.

Laevastu, T. and F. Favorite (1977) Preliminary report on dynamical numerical marine ecosystem model (DYNUMES II) for eastern Bering Sea. Processed report. Seattle, Northeast Fisheries Center, National Marine Fisheries Service.

Laevastu, T. and H. A. Larkins (1981) *Marine Fishery Ecosystem. Its Quantitative Evaluation and Management*. Fishing News: (Books), Guildfoul, 162 pp.

Lett, P. F. and A. C. Kohler (1976) Recruitment: a problem of multi-species interaction and environmental perturbation with special reference to the Gulf of St. Lawrence Atlantic herring (*Clupea harengus harengus*). *J. Fish. Res. Board Can.*, **33** (6): 1353–1371.

Mandecki, W. (1976) A computer simulation model for interacting fish population studies, applied to the Baltic. *Pol. Archiv. Hydrobiol.*, **23** (2): 281–308.

Marten, G. G. (1979) Impact of fishing on the inshore fishery of Lake Victoria (East Africa). *J. Fish. Res. Board Can.*, **36** (8): 891–900.

May, R. M. (1973) *Stability and Complexity in Model Ecosystems*. Princeton, New Jersey, Princeton University Press.

May, R. M. *et al.* (1979) Management of multispecies fisheries. *Science, Wash.*, **205** (4403): 267–277.

Munro, J. L. (ed.) (1974–6) The biology, ecology, exploitation and management of Caribbean reef fisheries. *Res. Rep. Zool. Dep. Univ. West Indies*, No. 3, Pts. 1–V1. 1: pp. var.

Murphy, G. I. (1966) Population biology of the Pacific sardine (*Sardinops caerulea*). *Proc. Calif. Acad. Sci.*, **34** (1): 1–84.

Nagasaki, F. (1973) Long-term and short-term fluctuations in the catches of pelagic fisheries around Japan. *J. Fish. Res. Board Can.*, **30** (12), Pt. 2: 2361–2367.

Paulik, G. J., A. S. Hourston and P. A. Larkin (1967) Exploitation of multiple stocks by a common fishery. *J. Fish. Res. Board Can.*, **24** (12): 2527–2537.

Pope, J. (1976) The effect of biological interaction on the theory of mixed fisheries. *Sel. Pap. ICNAF*, No. 1: 157–162.

Pope, J. (1979) *Stock assessment in Multi-species Fisheries*. Manila, FAO/UNDP South China Sea Development Programme, SCS/DEV/19: 106 pp.

Ricker, W. E. (1958) Maximum sustained yields from fluctuating environments and mixed stocks. *J. Fish. Res. Board Can.*, **15** (5): 991–1006.

Ryder, R. (1965) A method for estimating the potential fish production of north temperate lakes. *Trans. Am. Fish. Soc.*, **94**: 214–218.

Ryther, J. H. (1969) Photosynthesis and fish production in the sea. *Science, Wash.*, **166** (3901): 72–76.

Schaefer, M. B. (1965) The potential harvest of the sea. *Trans. Am. Fish. Soc.*, **94** (2): 123–128.

Turner, R. E. (1977) Intertidal vegetation and commercial yields of penaeid shrimp. *Trans. Am. Fish. Soc.*, **106** (5): 411–416.

Welcomme, R. L. (1976) Some general and theoretical consideration on the fish yield of African rivers. *J. Fish. Biol.*, **8**: 351–364.

Werner, E. E. (1979) Fisheries biology: Review of ecology of freshwater fish production *Science, Wash.*, **204**: 608–609.

Zuboy, J. R. and R. T. Lackey (1975) A computer simulation model of a multispecies centrarchid population complex. *Va. J. Sci.*, **26** (1): 13–19.

EXERCISES

Exercise 7.1

The average annual catch of haddock by trawlers fishing for herring in 1954–56 is given in Table 7.1

Table 7.1

Size	Weight caught (tons)	Weight landed (tons)	Fish (million)
Under 24 cm	2 500	—	25
24 to 27 cm	6 000	—	45
Over 27 cm	23 500	14 000	80
Total	32 000	14 000	150

The fishery for haddock lands on an average 70 000 tons per year of fish whose average weight is 330 g.

If the herring trawlers were to use a mesh of the minimum size for demersal fishing, what would be the effect on the haddock catches assuming:

(a) the legal mesh releases all haddock less than 24 cm;
(b) in the haddock fishery some 100 million fish over 24 cm are caught, but are rejected as being below the marketable size of 27 cm;
(c) the fish below 24 cm which would be released by a larger mesh take an average of six months to grow to 24 cm, during which time they would suffer a natural mortality of 0.2;
(d) $E = F/(F + M) = 0.8$ (where F includes all forms of fishing)?

Exercise 7.2

The declared management policy for 1979 for haddock on Georges Bank was to take 40 per cent of the stock at the beginning of the year, which was estimated to be 40 000 tons. Haddock are taken by trawlers directing their effort primarily at cod and silver hake. Quotas for these directed fisheries were 20 000 tons and 110 000 tons respectively.

(i) If the rules for incidental catches will allow, on the average, catches of haddock equal to 10 per cent and 5 per cent of the directed catches of cod and silver hake respectively, what should be the quota for the directed fishing for haddock?
(ii) If there is poor haddock recruitment, and the stock at the beginning of 1980 falls to 20 000 tons, what should be the 1980 quota for directed haddock fishing, if the other conditions and management objectives are unchanged?

[Assume:

(a) that the proportion of haddock in the catches of the other fisheries is unaltered; or
(b) that the relative fishing mortality on different species in each fishery is unaltered.

What do these assumptions imply about fishermen's behaviour? What do you think is more reasonable?]

Exercise 7.3

Table 7.2

Year	Sardines Spawners	Sardines Recruits	Anchovy Spawners	Year	Sardines Spawners	Sardines Recruits	Anchovy Spawners
1932	3627	6039		1945	684	1039	1400
1933	3291	1743		1946	409	1266	1400
1934	3221	1200		1947	391	2135	1500
1935	2912	2289		1948	525	1997	1600
1936	1707	2951		1949	684	503	1800
1937	1036	2876		1950	716	636	1950
1938	946	5021		1951	570	787	2150
1939	1032	5521		1952	554	674	2400
1940	1296	2563		1953	709	228	2700
1941	2001	1334	1350	1954	668	221	3000
1942	1947	1678	1360	1955	425	255	3400
1943	1678	1182	1370	1956	293	375	3800
1944	1174	758	1380	1957	211	260	4300

Table 7.2 (adapted from Murphy, 1966) gives data on the size of the spawning stock of California sardine, and of the resulting year-class, expressed as reproductive potential. Using the methods of Chapter 6, examine the relation between stock and recruitment separately for the periods 1932–48, and 1949–57.

(i) Does the Ricker function give an adequate description of the relation. If so, for each period: what spawning biomass gives the greatest recruitment? At what spawning biomass does the recruitment fall, on average, to a serious extent (take this as less than 80 per cent of the maximum)?

(ii) Do the data from the two periods fit the same stock–recruitment relation?

(iii) Table 7.2 also gives the estimated abundance of anchovy. Assuming that up to 1940 the abundance of anchovy was unchanged, examine the relation between recruitment of sardine and the abundance of anchovy. (Plot R/S for sardine against the total biomass of sardine plus anchovy.) Does this prove that the collapse of the sardine stock was caused by an increase in anchovy? (Bear in mind there appears to have been changes in growth and natural mortality of sardine.) Is the evidence good enough to suggest that control of the anchovy stock should be a part of any sardine management plan?

Answers to Exercises

Exercise 1.1. The c.p.u.e. increased greatly between 1971 and 1972, presumably due to increased efficiency. MSY is probably around 4600 tons, taken with 800–1000 boat months. Cost and value are about equal with effort at 1600 boat-months, catching 2900 tons of shrimp. MEY is about 1800 million rupiah (4300 tons of shrimp taken with an effort of 380 boat-months). Economic return at MSY is about 1500 million rupiah; (c) is the most desirable state.

Exercise 2.1 (i) Item (b) is stronger evidence of stock separation than (a). (ii) Recapture rates within each area were the same for each tagging position, strongly suggesting a single stock, with complete mixing taking about a year.

Exercise 2.2 (a)

	1973		1974		1975		1976	
	c.p.u.e.	%	c.p.u.e.	%	c.p.u.e.	%	c.p.u.e.	%
10–39	0.187	100	0.125	67	0.150	80	0.100	53
40–59	0.550	100	0.285	52	0.417	76	0.240	44
60–79	0.725	100	0.433	60	0.550	76	0.320	44
80–99	0.850	100	0.533	63	0.730	86	0.475	56
Total	0.470	100	0.335	71	0.495	105	0.340	72

(b)

	1973		1974		1975		1976	
Effort	1709		2090		2265		2463	
c.p.u.e.		0.550		0.320		0.437		0.276

(c) No; No.

(d)

	1973	1974	1975	1976
Fishing mortality	100	122	133	144
Abundance	100	58	79	50

Exercise 2.3 (a) 1970, 37; 1975, 58.
(b)

	1970	1975
Spring	26	48
Summer	26	50
Autumn	22	52
Winter	25	48

(c) 1970, 25; 1975, 49.
Dividing c.p.u.e. derived in (c) into the catch, the changes in effort are 1970 + 143%; 1975 + 55%.

Exercise 2.4. (a)

	Year 1	Year 2
Total effort	160	160
Catch (haddock)	626	1455
Catch (plaice)	479	203
c.p.u.e. (haddock)	3.91	9.09
c.p.u.e. (plaice)	2.99	1.27

(c) Haddock, mean c.p.u.e.: year 1, 5.00; year 2, 7.38; fishing intensity: year 1, 125; year 2, 196. Plaice. mean c.p.u.e.: year 1, 2.75; year 2, 1.62; fishing intensity: year 1, 174; year 2, 125. Total catch/total effort: haddock + 132%; plaice − 58%. Mean c.p.u.e: haddock + 49%; plaice − 41%. (Total catch divided by total effort overestimates the change in density.)

Exercise 2.5. **(a)**

	c.p.u.e.		Total effort		c.p.u.e. (as % of mean)	
	UK	USSR	UK	USSR	UK	USSR
1946	3059	1126	65	177	159	109
1947	3313	1020	10	334	172	99
1948	2612	979	155	416	136	95
1949	2831	949	171	511	147	92
1950	1468	841	243	423	76	81
1951	1305	821	313	497	68	79
1952	1272	1048	412	500	66	101
1953	1119	951	396	500	58	92
1954	1405	1190	425	502	73	115
1955	1508	1421	551	584	78	137
1956	1248	1041	630	756	65	101
1957	865	—	462	—	45	—
1958	831	—	467	—	43	—

The difference in trends in c.p.u.e. is probably because the UK data includes a factor to allow for increased fishing power, and probably gives a better measure of changes in abundance.

Exercise 2.6

	Cod	Redfish
Change in density	−3	−7
c.p.u.e. (all vessels)	−26	+82
c.p.u.e. cod (by redfish vessels)	+18	—
c.p.u.e. redfish (by cod vessels)	—	+29

Exercise 2.8 (i) pelagic 260000 tons; demersal 50000 tons. (ii) catfish 3000 tons; groupers 5000 tons; ponyfish 42000 tons; total 50000 tons. More ponyfish (84%) and fewer large fish. (iii) Pelagic 285000 tons; groupers 12500 tons. (This should not be treated as a firm estimate, but as an indication that the stock is already heavily fished, $F=0.37$)

Exercise 3.1 (i)

	1965	1966	1967	1968	1969	1970	1971	1972	1973	1975
c.p.u.e. (Indonesia)	—	—	—	—	0.163	0.099	0.082	0.114	0.177	0.155
c.p.u.e.(Malaysia)	0.028	0.036	0.055	0.048	0.032	0.021	0.021	0.019	0.030	0.030
Effort (Thailand)	—	42	214	423	1131	1887	1603	—	—	—

(ii) c.p.u.e. as % of 1969

Indonesia	—	—	—	—	100	61	50	70	109	95
Malaysia	88	113	172	150	100	66	66	59	94	94
Thailand	—	212	154	113	100	51	61	—	—	—
Mean	88	162	163	131	100	59	59	64	102	94

(iii) Thailand: Greatest sustained yield, under average conditions, 186000 tons with an effort of 1 300 000 hours. Malaysia: 110000 tons, at an effort of 4500 boats. Indonesia: Data are not good enough to determine yield curve.

Exercise 3.2

(i)			Total effort					
	Cod	Haddock		Cod	Haddock		Cod	Haddock
1906	108	64	1919	92	35	1929	470	186
1907	112	61	1920	137	76	1930	445	175
1908	104	57	1921	151	64	1931	436	190
1909	110	49	1922	173	66	1932	358	168
1910	121	54	1923	202	86	1933	398	186
1911	123	53	1924	275	127	1934	421	186
1912	133	60	1925	250	113	1935	345	193
1913	142	68	1926	271	116	1936	246	179
			1927	343	142	1937	212	227
			1928	409	168	1938	275	150
1946	129	44	1951	348	294	1956	473	326
1947	168	127	1952	431	250	1957	559	475
1948	231	163	1953	464	294	1958	639	500
1949	269	178	1954	541	365	1959	668	356
1950	328	250	1955	483	305			

(ii) There is little suggested that for cod, different relations between effort and c.p.u.e. hold for different periods. For haddock it seems that after 1950 c.p.u.e. is, for a given effort, higher than before 1950. (iii) (a) Moderate increase; (b) large reduction (probably to less than half); (c) probably a moderate reduction (the maximum is not well determined, at least graphically).

Exercise 3.3 The correlation between abundance and effort is not significant. It is therefore possible that fishing is having no effect on the stock, though the confidence limits are such that it could be having an effect. If effort is turned to skipjack when it is more abundant than yellowfin, then a positive correlation would be expected. In its absence the belief that fishing is having an effect is strengthened.

Exercise 3.4 (i) By adding the catches between 1939 and 1946 to the changes in population between 1939 and 1946 to estimate the average annual sustainable yield during this period. (ii) If c.p.u.e. of $1 = 20000$ whales; $\text{MSY} = 11\text{–}13000$ whales at a population of 90–100000 whales. If c.p.u.e. of $1 = 30000$ whales; $\text{MSY} = 11\text{–}13000$ whales at a population of 140–160000 whales.

Note: c.p.u.e. before 1932/3 is probably unreliable, but unexploited population (with zero sustainable yield) is not more than 75000 whales (the pre-1932 catch) greater than the population in 1932/3.

Exercise 4.1 Two age-groups can be clearly distinguished (modes at 10.6 and 14.9 cm) and one less clearly (at 19.4 cm). The numbers in the groups are approximately 5710, 4600, and 2040.

Exercise 4.2. (i) Two strong groups and one weak group can be followed for most of their life, and two weak groups for part of their life. (ii) All years except 1964 are consistent. (iii) About 12 years.

Exercise 4.3 $L_\infty = 37.5$, $K = 2.3$ (estimated graphically), $t_0 = 1.85$ (mean ages 3–12).

Predicted lengths (ages 3–15) 25.4, 27.9, 29.9, 31.5, 32.7, 33.7, 34.5, 35.1, 35.6, 36.0, 36.3, 36.6, 36.7. Predicted weights: 128, 169, 208, 243, 274, 299, 320, 338, 353, 364, 374, 381, 385.

Exercise 4.4 (i) $L_\infty = 162$, $K = 0.46$. (ii) Sizes at ages 1–6: 59.4, 97.0, 120.8, 135.8, 145.3, 151.3. (Note that the data give growth increment for less than one year.)

Exercise 4.5 (i) Jan 6.6 cm, Feb 7.5 cm, Mar 9.5 cm, Apr 10.3 cm and 6.4 cm, May 10.4 and 7.5 cm. (ii) $L_\infty = 10.5$ cm, $K =$ (a) 1.1 (on monthly basis) (b) 13.2 (on a yearly basis) Results may be unreliable because of seasonal changes in growth.

Exercise 4.6 (a) 56.3%, 86.6%, 42.2%, 0.288; 81.0%, 94.9%, 72.9%, 0.105; 1.0%, 31.6%, 0.10%, 2.303; 25.0%, 70.7%, 12.5%, 0.693. (b) 0.5; 0.8; 1.0. (c) No. 44%; 79%; 94%. (d) Yes (i) − 0.23, −0.23, 0.79, − 0.23; (ii) −0.695, −0.34, 0.71 −695. Yes, the data fit a constant mortality rate (i) survival = 79% mortality coefficient = 0.23; (ii) survival = 50% mortality coefficient = 0.695.

Exercise 4.7 Possible estimates are: II–III, 0.81; III–IV, 0.88; IV–V, 0.85; V+–VI+, 0.83; II+–III+, 0.83. Plotting the data for the first set of hauls as a catch curve gives an estimate of about 0.95.

Exercise 4.8 674; 294; 0.829.

Exercise 4.9 (i) From five years old. (ii) 1929–38, $Z = 0.82$; 1950–58, $Z = 0.55$. (iii) $M = 0.118$, $F_1 = 0.70$, $F_2 = 0.43$.

Exercise 4.10

	Ages						
Years	5/6	6/7	7/8	8/9	9/10	10/11	8+/7+
1932/33	−0.81	−0.38	0.13	0.30	0.13	1.24	0.42
1933/34	−1.07	−0.49	−0.63	0.26	0.76	0.89	−0.05
1934/35	−1.48	−0.49	−0.06	1.12	1.08	0.95	0.60
1935/36	−0.09	0.57	0.08	0.51	1.11	1.53	0.32
1936/37	−1.48	−0.69	−0.09	1.30	2.28	1.28	0.64
1937/38	−0.67	−0.21	0.58	0.70	0.69	−0.24	0.62
1946/47	−0.92	−0.18	0.83	0.51	0.31	0.47	0.52
1947/48	−0.15	0.06	0.69	−0.08	1.26	1.06	0.83
1948/49	−0.80	0.59	1.08	0.56	0.89	0.24	0.86
1949/50	1.21	0.99	0.77	0.66	1.31	1.07	0.94
1950/51	−2.47	−0.74	0.67	0.82	0.88	−0.10	0.72
1951/52	−0.64	0.03	1.03	1.14	0.99	0.66	1.00
1952/53	−0.33	0.82	0.88	0.93	1.11	2.32	1.02
1953/54	−0.46	−0.21	0.59	1.22	1.21	−0.04	0.80
1954/55	−0.34	0.33	1.32	0.72	0.22	−0.17	1.06
1955/56	−0.75	0.69	1.03	1.51	1.15	1.69	1.18
1956/57	−0.44	0.86	0.78	1.47	0.72	0.48	0.87
1957/58	−0.90	0.52	0.56	1.44	0.34	0.33	0.77

$M = 0.25$ F (1958) = 0.60, approximately.

Exercise 4.11

(a)				Values of fishing morality					
Age	1956	1957	1958	1959	1960	1961	1962	1963	1964
3	0.01	0.02	0.01	0.01	0.01	0.02	0.01	0.02	0.02
4	0.02	0.04	0.04	0.05	0.04	0.08	0.11	0.11	0.09
5	0.09	0.23	0.08	0.15	0.13	0.18	0.35	0.28	0.27
6	0.16	0.23	0.19	0.21	0.18	0.30	0.43	0.44	0.47
7	0.30	0.22	0.29	0.28	0.23	0.45	0.55	0.70	0.55
8	0.22	0.30	0.36	0.24	0.31	0.44	0.70	0.61	0.96
9	0.29	0.21	0.42	0.20	0.27	0.55	0.52	0.55	1.03
10	0.24	0.16	0.29	0.26	0.27	0.39	0.69	0.52	0.76
11	0.40	0.20	0.37	0.18	0.30	0.40	0.56	0.79	0.70
12	0.34	0.33	0.43	0.20	0.27	0.33	0.40	0.97	1.04

Note: The method is not suitable for estimating the values of fishing mortality in 1963 and 1964, or for ages above 12.

Exercise 4.12 $Z = 0.53$, $F = 0.098$ (units of 10 days) $Z = 19.35$, $F = 3.58$ (annual units).

Exercise 4.13 (i) No. (ii) Difference exposure to predators; difference chances of recaptured shrimp being observed and returned. (iii) 1975: 765; 1976: 870. (iv) Differences may be due to local differences in fishing intensity.

Exercise 4.14 (i) Higher returns from night than day. (ii) Increased predation when shrimp are easily visible (day, in clear water). (iii) Feb 700; Aug 850; Oct 644.

Exercise 4.15

	Tags returned per 1000 tons	Number recaptured
Canada	6.28	392
France	1.14	176
Portugal	5.00	36
Spain	1.85	51

Exercise 4.16 Here, $q = 14$ per 100 hours' fishing per square per week.

Exercise 4.17 60%; $F = 1.03$, $M = 0.17$.

Exercise 4.18

Length (cm)	65 –	70 –	75 –	80 –	85 –	90 –
Ratio gill-nett: purse-seine	0	0.5	0.625	1.154	1.500	1.786
Ratio, as % of 90–95 cm	0	28	35	65	84	100

Length (cm)	95 –	100 –	105 –	110 –	115 –	120 +
Ratio gill-net: purse-seine	1.286	0.794	0.600	0.333	0.250	0.200
Ratio, as % of 90–95 cm	72	44	34	19	14	11

Exercise 4.19 (a) $l_c = 24.7$ cm, s.f. = 2.21; (b) $l_c = 23.9$ cm, s.f. = 3.23.

Exercise 4.20

	Age					
Mesh size	2	3	4	5	6	– 7
8.0	0.83	0.99	0.99	1.00	1.00	1.00
10.0	0.54	0.73	0.97	0.99	1.00	1.00
11.4	0.13	0.27	0.73	0.84	0.94	0.97

Exercise 5.1 (a) 556 tons; (b) 908 tons; (c) 780 tons. The assumption of a constant percentage dying of natural causes will overestimate natural deaths. The highest fishing rate (40% caught), corresponds closely to mortalities $F = 0.55$, and $M = 0.1$. At these rates about 7% of those alive at the beginning of a period will die of natural mortality during that period.

Exercise 5.2 Here, $l_c = 24.8$, $c = 0.36$. (i) (a) Y/R decreases by 9%; (b) Y/R increases by 15%. (ii) (a) c.p.u.e. decreases by 32%; (b) c.p.u.e. increases by 130%. (iii) A decrease of about 60%. (iv) Max Y/R occurs at $c = 0.68$, $l_c = 46.6$ in mesh size of 212 mm, and will be some 57% greater than the present.

Exercise 5.3

	M/K			
	0.5	0.75	1.0	1.5
(a) Change in Y/R by increasing F by 50%	– 13%	– 11%	– 8%	– 1%
(b) Change in Y/R by decreasing F by 30%	+ 14%	+ 9%	+ 5%	– 22%
(c) % change in F to give max. Y/R	– 80%	– 65%	– 50%	+ 5%
(d) Size at first capture giving max. Y/R	45 cm	41 cm	37 cm	31 cm
Mesh size giving max. Y/R	141 mm	129 mm	117 mm	98 mm

Exercise 5.4 (a) Decrease by 11.6%. (b) Number of released fish ultimately caught = 617 000; long-term effect on catches—no change. (c) Total catch up 325 tons (3%); line catch up 533 tons (9%); trawl catch down 208 tons (4%). (d) Total catch up 1927 tons (17%); line catch up 1408 tons (23%); trawl catch up 519 tons (9%).

Exercise 5.5 (c)

Value of M	0.4	0.6	0.8	1.2
F of new fishery	0.55	0.45	0.35	0.15
Total F	1.65	1.35	1.05	0.45
Catches '000 tons, total	140.7	151.6	163.3	194.6
Existing Vessels	93.8	101.1	108.9	129.7
New vessels	46.9	50.5	54.4	64.9

Exercise 6.1

	1968	1969	1970	1971	1972	1973	1974	Total
Light fishing	209.5	238.5	294	395.5	425	272.5	345	2180
Moderate fishing	157.5	424	540.5	459	470	325	485	28610
Heavy fishing	71	791.5	448.5	460.5	445	247.5	740	3194

Greatest total yield is taken with heavy fishing, but greatest yield in worst year (1968 in all strategies) is taken with light fishing. (a) Heavy fishing; (b) light fishing.

Exercise 6.2

	(i) Growth increments (cm)			(ii) Fitted growth curve		
	2nd year	3rd year	4th year	K	L_∞	L_∞ if $K = 0.35$
1927	7.5	5.2	—	0.40	38.5	40.6
1928	8.5	4.2	3.8	0.48	36.7	39.0
1929	7.5	4.4	4.0	0.35	38.9	38.9
1930	6.8	4.2	3.3	0.34	37.8	37.2
1931	7.4	5.1	4.1	0.43	37.8	40.8
1932	7.3	4.9	3.1	0.49	36.6	40.1
1933	7.3	5.2	5.3	0.17	58.6	41.8
1934	9.0	5.5	3.7	0.56	38.6	42.4
1935	8.9	5.6	3.5	0.51	41.8	45.4

There appears to be a clear association between increasing abundance (as measured by the c.p.u.e.), and decreasing L_∞, if K is assumed fixed at 0.35.

Exercise 6.3 (i) (1) A decrease of M to a little more than 0.07 (i.e., 30%); (2) an increase of M to a little over 0.15 (i.e., 50%).

Exercise 6.4 (i) Adult stock 96 units, expected recruitment, 8.35 units. (ii) For stock A, average recruitment does not appear to change with changes in stock. Advice would be based on the value yield-per-recruit and other considerations. For stock B (a) stock size is about optimal, do not allow it to change much; (b) stock size is somewhat too great, and should be decreased; (c) stock size is seriously too low, and rebuilding the spawning stock should receive high priority. (iii) Recruitment may start falling at stock sizes below 1, or thereabouts. If current stock is 2 units it would be unwise to allow it to decrease further.

Exercise 6.5 (1) (b) Using a division of both stock and recruitment into thirds (large, moderate, and small), to give 3 × 3 tables, the χ^2 values are: haddock 11.14, significant at 0.05 level; mackerel 11.08, significant at 0.05 level; shad 6.35, not significant; salmon 3.42, not significant. The decline in mackerel recruitment, since 1962, precedes the decline in stock. The four years (1965–68) of small recruitment and small stock probably reflects the influence of recruitment on stock, not vice versa. (2) If the Ricker curve is fitted by a least-squares regression of $\log S/R$ on S (this is not necessarily the only method) the parameters in the equation $R = aSc^{-bS}$ are: haddock $a = 0.718$, $b = 0.00827$; mackerel $a = 0.659$, $b = 0.00217$; shad $a = 0.740$, $b = 0.00508$; salmon $a = 1.084$, $b = 0.00658$.

Exercise 6.6 Current equilibrium is approximately $S = 80$, $R = 85$. (a) (i) Equilibrium $S = 102$, $R = 90$; (ii) equilibrium $S = 53$, $R = 70$; (iii) no clear equilibrium. (b) (i) $S = 96$, $R = 85$; (ii) $S = 64$, $R = 85$; (iii) $S = 40$, $R = 85$. Since increasing fishing could decrease both yield-per-recruit, and recruitment, this should be prevented, and measures should be taken to reduce fishing. (c) (i) $S = 96$, $R = 86$; (ii) $S = 54$, $R = 73$; (iii) No equilibrium, stock would collapse.
Note: answer to (c) can vary depending on how the curve is drawn.

Exercise 7.1 Total numbers caught above 24 cm = 437 million; Number released = 25 million, of which 18.1 million will be subsequently caught. Catches will therefore increase by 4.1%, or 2900 tons in haddock fishery, and 580 tons in the herring trawl fishery.

Exercise 7.2 (i) Total allowable catch 16 000 tons, 8500 tons in directed fishery. (ii) (a) TAC 8000 tons, 500 tons in directed fishery; (b) TAC 8000 tons, 4250 tons in directed fishery. (a) Assumes those fishing primarily for cod or silver hake will change their fishery strategy so as to increase effective mortality on haddock, and thus maintain the absolute quantity of haddock caught. (b) Assumes that the fishing practice is unchanged. Either could be reasonable, (b) if those fishing other species are not really interested in haddock, (a) if they do try to maintain a significant haddock catch.

Exercise 7.3 (i) For period 1932–48, a Ricker curve describes the relation well. Maximum recruitment at a stock of 1280 units; 80% of maximum at 600 units. For period 1949–57, the fit to a Ricker curve is less good. If one is fitted (by linear regression of $\ln R/S$ on S,) the maximum recruitment occurs at a stock of 600 units. (ii) The data for the periods do not fit the same relations. (iii) The data for the two periods fit, approximately, the same relation between R/S (of sardines) and total spawning biomass (sardine plus anchovy). This does not prove the increase of anchovy was responsible for the decline of sardine, but strongly suggests that control of the anchovy should be part of the sardine management plan.

Appendix I

Tables of yield per recruit for selected values of M/K. For complete tables, and a description of the approach the reader is referred to the original tabulations (Beverton and Holt, 1964). In the tables, the entry underlined gives the maximum value of Y/R for a given value of $E(=F/(F+M)$, i.e. amount of fishing), the size of first capture (i.e., $c=l_0/L_{00}$) being varied. The asterisked entry gives the maximum value of Y/R for a given c, E being varied.

REFERENCE

Beverton, R. J. H. and S. J. Holt. (1964). Tables of yield function for fishery assessment. *FAO Fish. Tech. Paper* 38, 49 p.

(a) Yield per nominal recruit, $M/K = 0{,}50$

E	0,05	0,10	0,15	0,20	0,25	0,30	0,35	0,40	0,45	0,50
c										
98	006927	013843	020748	027640	034517	041379	048223	055046	061845	068618
96	009596	019164	028699	038199	047659	057075	066440	075748	084991	094158
94	011514	022977	034382	045722	056991	068179	079276	090270	101145	111886
92	013027	025976	038837	051600	064255	076788	089183	101421	113482	125338
90	014272	028436	042479	056389	070149	083741	097143	110331	123275	135938
88	015322	030503	045529	060381	075040	089481	103678	117598	131203	144447
86	016220	032266	048119	063757	079155	094284	109110	123593	137686	151334
84	016997	033785	050340	066637	082646	098331	113652	128563	143006	156915
82	017674	035102	052257	069108	085621	101754	117462	132687	147366	161420
80	018267	036249	053916	071234	088162	104653	120654	136101	150920	165022
78	018787	037251	055357	073066	090333	107105	123321	138910	153787	167855
76	019245	038127	056608	074644	092185	109171	125535	141198	156066	170030
74	019648	038894	057694	076001	093759	110902	127356	143034	157833	171635
72	020004	039565	058636	077165	095089	112340	128834	144476	159157	172744
70	020317	040151	059450	078157	096206	113520	130010	145574	160092	173423
68	020593	040662	060151	078999	097134	114473	130922	146369	160687	173724
66	020836	041106	060752	079708	097895	115226	131600	146898	160983	173696
64	021049	041491	061264	080297	098508	115802	132072	147192	161018	173381
62	021235	041823	061697	080781	098989	116220	132362	147280	160824	172814
60	021398	042107	062059	081171	099353	116500	132491	147187	160428	172028
58	021539	042350	062357	081477	099614	116656	132479	146935	159857	171052
56	021661	042554	062599	081709	099783	116704	132342	146543	159134	169912
54	021767	042724	062791	081875	099870	116656	132095	146030	158278	168631
52	021856	042864	062939	081982	099886	116524	131754	145411	157308	167230
50	021933	042978	063046	082038	099839	116318	131329	144701	156241	165728
48	021997	043067	063119	082048	099736	116048	130832	143914	155092	164142
46	022050	043135	063161	082018	099585	115723	130274	143060	153875	162487
44	022094	043185	063175	081952	099391	115349	129664	142151	152601	160778
42	022129	043218	063165	081857	099162	114935	129009	141196	151282	159026
40	022156	043236	063135	081734	098902	114487	128318	140205	149928	157243
38	022177	043242	063086	081590	098616	114010	127598	139185	148548	155440
36	022192	043236	063022	081426	098307	113509	126855	138143	147150	153626
34	022202	043222	062945	081246	097982	112991	126093	136086	145742	151808
32	022208	043199	062857	081053	097641	112458	125320	136020	144331	149995
30	022210	043169	062760	080850	097290	111915	124538	134951	142921	148193
28	022209	043134	062655	080638	096931	111365	123753	133882	141519	146408
26	022205	043094	062545	080420	096566	110812	122967	132818	140130	144646
24	022199	043050	062430	080197	096198	110258	122184	131762	138756	142909
22	022191	043004	062311	079972	095829	109705	121407	130718	137403	141203
20	022182	042955	062190	079745	095460	109156	120638	129689	136072	139531
18	022171	042904	062069	079519	095093	108613	119880	128677	134766	*137894
16	022160	042853	061946	079293	094730	108077	119134	127683	133488	*136297
14	022148	042801	061824	079069	094371	107549	118401	126710	132240	*134739
12	022136	042749	061703	078848	094018	107030	117683	125759	131021	*133223
10	022124	042697	061583	078631	093672	106522	116981	124830	129834	*131748
00	022063	042450	061016	077605	092045	104150	113716	120530	124367	*125000
F/M	0,0526	0,111	0,177	0,250	0,333	0,429	0,539	0,667	0,818	1,00
F/K	0,0260	0,056	0,088	0,130	0,170	0,210	0,270	0,330	0,410	0,50

0,55	0,60	0,65	0,70	0,75	0,80	0,85	0.90	0,95	1,00	E
										c
075359	082063	088725	095336	1011887	108363	114750	121024	127156	*133105	98
103239	112219	121081	129802	138356	146708	154812	162610	170023	*176947	96
122471	132874	143064	153003	162642	171921	180760	189057	196679	*203451	94
136959	148306	159333	169983	180184	189845	198852	207056	214268	*220246	92
148280	160247	171778	182794	193199	202873	211663	219375	225764	*230530	90
157277	169625	181412	192539	202886	212300	220591	227522	232797	*236069	88
164469	177012	188864	199907	209992	218938	226518	232453	236404	*237990	86
170212	182800	194565	205365	215028	223343	230049	234827	*237305	237082	84
174759	187271	198824	209255	218369	225924	231628	235129	*236029	233926	82
178303	190638	201877	211837	220297	226991	231596	*233733	232986	228973	80
180997	193071	203911	213314	221038	226790	230222	*230934	228502	222585	78
182961	194704	205075	213853	220774	225522	*227727	226974	222845	215053	76
184296	195649	205493	213589	219654	223351	*224293	222055	216234	206625	74
185086	195999	205268	212638	217806	*220418	220073	216344	208858	197504	72
815402	195834	204490	211097	215337	*216841	215199	209988	200875	187869	70
185306	195224	203234	209051	212341	*212724	209783	203111	192425	177870	68
184850	194226	201567	206573	*208899	208156	203926	195823	183628	167638	66
184083	192894	199546	203727	*205082	203215	197714	188219	174587	157286	64
183044	191273	197223	200571	*200953	197971	191223	180383	165395	146915	62
181770	189405	194643	*197154	196567	192486	184521	172389	156132	136610	60
180294	187325	191847	*193521	191975	186813	177667	164303	146871	126447	58
178645	185067	188870	*189714	187220	181003	170716	156183	137673	116490	56
176849	182659	185747	*185767	182342	175097	163715	148081	128594	106797	54
174930	180127	*182505	181712	177378	169137	156706	140043	119682	097416	52
172909	177497	*179171	177580	172357	163155	149728	132109	110980	088388	50
170804	174788	*175769	173396	167310	157184	142813	124315	102524	079749	48
168634	172020	*172319	169183	162262	151251	135992	116692	094346	071527	46
166413	*169210	168842	164963	157236	145381	129292	109268	086474	063746	44
164157	*166375	165354	160755	152252	139595	122734	102067	078931	056424	42
161877	*163528	161871	156574	147328	133915	116341	095109	071734	049574	40
159585	*160681	158407	152437	142481	128355	110128	088412	064900	043206	38
157292	*157847	154973	148357	137725	122932	104112	081989	058441	037325	36
155006	*155035	151582	144345	133073	117658	098305	075854	052365	031931	34
*152736	152254	148243	140411	128534	112544	092718	070013	046677	027021	32
*150489	149512	144963	136566	124119	107600	087358	064475	041381	022590	30
*148272	146817	141752	132815	119834	102831	082233	059244	036476	018627	28
*146089	144173	138614	129167	115686	098245	077347	054321	031959	015119	26
*143947	141586	135555	125625	111681	093846	072704	049708	027826	012051	24
*141848	139061	132579	122194	107822	089636	068303	045402	024069	009404	22
*139797	136600	129690	118878	104111	085617	064145	041399	020678	007155	20
137796	134208	126891	115672	100551	081790	060229	037695	017641	005281	18
135847	131884	124182	112595	097140	078152	056655	034281	014945	003754	16
133953	129632	121566	109631	093880	074703	053104	031150	012573	002545	14
132113	127451	119042	106784	090767	071439	049885	028290	010508	001621	12
130328	125342	116610	104053	087801	068354	046885	025691	008729	000949	10
122216	115837	105764	092045	075000	055411	034843	016071	003322	000000	00
1,22	1,50	1,86	2,33	3,00	4,00	5,67	9,00	19,00	∞	F/M
0,61	0,75	0,93	1,20	1,50	2,00	2,90	4,50	9,50	∞	F/K

(b) Yield per nominal recruit $M/K = 1.00$

E	0,05	0,10	0,15	0,20	0,25	0,30	0,35	0,40	0,45	0,50
c										
98	000051	000102	000153	000204	000255	000306	000356	000407	000457	000507
96	000165	000330	000495	000658	000822	000984	001146	001307	001468	001627
94	000323	000645	000965	001283	001600	001914	002227	002538	002846	003152
92	000514	001024	001531	002034	002533	003029	003520	004007	004489	004966
90	000729	001452	002168	002879	003582	004278	004967	005647	006319	006982
88	000692	001915	002859	003791	004713	005623	006521	007405	008276	009133
86	001209	002405	003586	004751	005899	007031	008144	009238	010311	011363
84	001466	002911	004336	005740	007120	008475	009805	011108	012383	013627
82	001727	003427	005100	006742	008354	009933	011478	012986	014456	015886
80	001991	003946	005866	007746	009587	011385	013138	014844	016501	018105
78	002254	004463	006626	008741	010804	012814	014767	016661	018493	020259
76	002514	004973	007375	009716	011994	014206	016349	018418	020411	022323
74	002769	005471	008104	010664	013147	015550	017869	020100	022238	024279
72	003258	005956	008811	011578	014255	016835	019316	021692	023958	026110
70	003258	006423	009489	012453	015310	018054	020681	023186	025562	027804
68	003489	006870	010136	013284	016306	019199	021956	024572	027039	029352
66	003711	007296	010749	014067	017241	020266	023136	025843	028383	030746
64	003921	007698	011326	014799	018109	021250	024215	026997	029588	031981
62	004119	008077	011865	015479	018909	022149	025192	028029	030652	033055
60	004306	008430	012365	016104	019639	022962	026064	028937	031574	033967
58	004480	008757	012825	016675	020298	023686	026831	029723	032354	034717
56	004641	009059	013245	017191	020887	024324	027493	030386	032993	035307
54	004790	009335	013626	017652	021405	024875	028052	030928	033494	035742
52	004927	009586	013966	018059	021854	025341	028511	031354	033862	036027
50	005051	009811	014268	018413	022235	025724	028871	031666	034100	036167
48	005164	010011	014532	018716	022551	026028	029137	031868	034215	036169
46	005264	010188	014760	018969	022804	026256	029313	031968	034212	036041
44	005354	010342	014953	019175	022997	026410	029410	031969	034100	035792
42	005432	010474	015112	019336	023134	026496	029414	031879	033886	035430
40	005501	010585	015240	019454	023217	026518	029350	031704	033577	034966
38	005560	010676	015338	019533	023250	026481	029217	031452	033182	034408
36	005609	010749	015408	019575	023238	026389	029021	031129	032710	033768
34	005651	010806	015453	019583	023183	026247	028768	030742	032170	033055
32	005684	010846	015475	019560	023091	026061	028465	030301	031570	032280
30	005710	010873	015476	019510	022965	025835	028117	029811	030919	*031453
28	005730	010887	015458	019435	022809	025575	027732	029280	030227	*030585
26	005745	010889	015424	019339	022627	025286	027314	028716	029500	*029684
24	005754	010882	015375	019224	022424	024973	026871	028126	028749	*028761
22	005758	010866	015313	019094	022203	024640	026408	027516	*027981	027826
20	005759	010843	015242	018951	021967	024292	025930	026894	*027203	026886
18	005757	010814	015162	018799	021721	023933	025443	026265	*026423	025950
16	005752	010780	015076	018639	021468	023568	024951	025636	*025648	025025
14	005745	010742	014986	018474	021210	023200	024460	*025010	024882	024118
12	005737	010702	014892	018306	020950	022833	023973	*024394	024133	023235
10	005727	010660	014796	018138	020692	022470	023493	*023791	023403	022381
00	005674	010448	014330	017331	019471	020779	*021293	021064	020156	018648
F/M	0,0526	0,111	0,177	0,250	0,333	0,429	0,539	0,667	0,818	1,00
F/K	0,0920	0,190	0,310	0,440	0,580	0,750	0,940	1,200	1,400	1,80

0,55	0,60	0,65	0,70	0,75	0,80	0,85	0.90	0,95	1,00	E
										c
000558	000608	000657	000707	000757	000855	000855	000904	000953	*001001	98
001786	001944	002101	002257	002411	002565	002717	002868	003018	*003165	96
003456	003757	004055	004350	004641	004929	005214	005494	005770	*006042	94
005437	005903	006362	006816	007262	007700	008131	008554	008967	*009371	92
007635	008278	008910	009530	010138	010733	011314	011880	012430	*012964	90
009974	010798	011606	012395	013164	013912	014639	015342	016021	*016673	88
012393	013397	014377	015329	016253	017146	018008	018835	019627	*020381	86
014840	016019	017162	018268	019334	020358	021338	022272	023157	*023157	84
017273	018615	019910	021115	022347	023485	024564	025582	026537	*027426	82
019655	021147	022578	023945	025244	026474	027630	028709	029708	*030625	80
021956	023581	025130	026599	027985	029283	030491	031604	032620	*033537	78
024151	025890	027537	029086	030535	031878	033111	034232	035238	*036125	76
026218	028052	029775	031382	032869	034231	035464	036565	037531	*038362	74
028142	030048	031825	033465	034966	036321	037527	038581	039481	*040228	72
029907	031865	033672	035322	036811	038133	039286	040266	041074	*041712	70
031505	033492	035306	036941	038393	039658	040732	041614	042305	*042809	68
032928	034921	036718	038316	039707	040890	041861	042620	043172	*043523	66
034170	036147	037906	039442	040750	041827	042672	043288	043680	*043860	64
035230	037170	038868	040320	041522	042472	043171	043624	*043839	043833	62
036107	037989	039606	040953	042028	042832	043366	043638	*043662	043457	60
036803	038606	040122	041345	042275	042913	043266	*043344	043164	042753	58
037321	039027	040422	041504	042271	042729	*042886	042758	042367	041745	56
037665	039257	040515	041438	042028	*042292	042243	041900	041292	040458	54
037842	039304	040410	041160	041559	*041618	041354	040791	039964	038921	52
037860	039177	040116	040681	*040879	040725	040239	039454	038410	037163	50
037726	038885	039646	*040015	040004	039630	038921	037913	036656	035215	48
037450	038439	039013	*039178	038951	038354	037420	036194	034733	033110	46
037042	037853	*038230	038185	037739	036918	035762	034324	032669	030881	44
036512	037136	*037311	037053	036386	035343	033971	032329	030495	028559	42
035873	*036304	036273	035799	034911	033651	032070	030238	028240	026178	40
035134	*035369	035129	034440	033336	031863	030084	028076	025935	023770	38
034308	*034344	033896	032994	031678	030003	028038	025871	023607	021366	36
*033408	033244	032589	031478	029959	028092	025957	023650	021286	018995	34
*032444	032082	031224	029910	028196	026151	023862	021437	018999	016686	32
031429	030871	029815	028307	026410	024201	021778	019256	016770	014464	30
030374	029625	028378	026686	024618	022262	019725	017131	014624	012354	28
029291	028357	026927	025061	022839	020354	017723	015083	012583	010377	26
028191	027078	025475	023450	021087	018493	015793	013131	010666	008552	24
027084	025801	024036	021864	019380	016696	013950	011292	008889	006893	22
025980	024536	022622	020319	017730	014979	012209	009582	007267	005414	20
024887	023294	021243	018825	016151	013353	010584	008013	005811	004121	18
023816	022084	019910	017393	014653	011830	009086	006595	004528	003019	16
022772	020914	018632	016033	013245	010418	007722	005333	003422	002107	14
021762	019790	017415	014751	011935	009126	006498	004232	002494	001382	12
020792	018719	016265	013553	010728	007955	005416	003292	001739	000832	10
016639	014246	011607	008883	006250	003898	002009	000730	000113	000000	00
1,22	1,50	1,86	2,33	3,00	4,00	5,67	9,00	19,00	∞	F/M
2,10	2,60	3,20	4,10	5,30	7,00	9,90	16,00	33,00	∞	F/K

(c) Yield per nominal recruit $M/K = 1.75$

E	0,05	0,10	0,15	0,20	0,25	0,30	0,35	0,40	0,45	0,50
c										
98	000970	001938	002904	003869	004831	005791	006749	007704	008657	009606
96	001880	003754	005621	007481	009333	011177	013011	014835	016647	018447
94	002734	005454	008160	010849	013521	016174	018806	021416	024001	026559
92	003534	007043	010527	013982	017408	020800	024156	027473	030748	033976
90	004281	008525	012730	016892	021007	025071	029081	033031	036916	040730
88	004979	009906	014777	019587	024331	029004	033601	038113	042534	046855
86	005629	011189	016674	022077	027394	032615	037734	042742	047628	052381
84	006234	012379	018428	024373	030207	035920	041501	046941	052225	057341
82	006796	013481	020047	026484	032783	038932	044920	050731	056352	061764
80	007316	014498	021536	028419	035134	041669	048008	054135	060032	066680
78	007797	015435	022903	030186	037271	044143	050783	057173	063293	069117
76	008241	016297	024153	031795	039206	046369	053263	059866	066156	072104
74	008650	017086	025293	033254	040950	048361	055463	062234	068645	074668
72	009025	017807	026329	034571	042513	050131	057401	064296	070784	076835
70	009368	018463	027266	035754	043905	051694	059092	066070	072595	078630
68	009682	019059	028110	036812	045138	053061	060552	067575	074097	080079
66	009967	019597	028867	037751	046220	054246	061794	068829	075314	081205
64	010226	020082	029542	038579	047162	055259	062834	069849	076264	082033
62	010460	020516	030140	039303	047972	056112	063685	070651	076966	082584
60	010670	020902	030666	039930	048659	056817	064362	071252	077441	082880
58	010859	021244	031125	040467	049233	057383	064876	071666	077705	082943
56	011027	021545	031522	040920	049701	057823	065241	071910	077777	082792
54	011176	021808	031861	041295	050071	058144	065469	071996	077673	082447
52	011308	022035	032146	041599	050351	058358	065571	071939	077410	981927
50	011423	022230	032382	041836	050549	058474	065559	071753	077002	081250
48	011523	022394	032572	042014	050672	058499	065443	071450	076465	080433
46	011610	022531	032721	042136	050727	058444	065234	071042	075814	079492
44	011683	022642	032833	042208	050720	058315	064941	070541	075060	078443
42	011745	022731	032910	042236	050657	058121	064574	069958	074219	077301
40	011797	022798	032956	042223	050545	057870	064142	069304	073301	076080
38	011839	022847	032975	042173	050389	057567	063652	068588	072319	074793
36	011873	022879	032969	042092	050195	057221	063115	067820	071283	073454
34	011899	022896	032942	041984	049967	056837	062536	067009	070204	072073
32	011918	022900	032895	041851	049711	056420	061923	066164	069091	070661
30	011931	022892	032832	041697	049431	055978	061282	065291	067954	*069230
28	011939	022875	032756	041526	049131	055514	060621	064399	066801	*067788
26	011942	022849	032667	041341	048815	055034	059944	063493	065639	*066345
24	011942	022816	032569	041144	048487	054542	059256	062581	064475	*064908
22	011939	022778	032463	040938	048150	054042	058564	061668	063317	*063484
20	011933	022734	032351	040726	047807	053538	057870	060758	*062169	062080
18	011924	022687	032234	040510	047460	053033	057179	059857	*061037	060702
16	011914	022637	032115	040291	047113	052530	056495	058969	*059925	059354
14	011903	022586	031994	040072	046768	052032	055820	058096	*058837	058040
12	011891	022533	031872	039853	046425	051540	055156	057241	*057777	056765
10	011878	022480	031750	039637	046088	051058	054507	056408	*056746	055530
00	011814	022221	031170	038612	044505	048814	051514	*052597	052078	050000
F/M	0,0526	0,111	0,177	0,250	0,333	0,429	0,539	0,667	0,818	1,00
F/K	0,0530	0,110	0,180	0,250	0,330	0,430	0,540	0,670	0,820	1,00

0,55	0,60	0,65	0,70	0,75	0,80	0,85	0.90	0,95	1,00	E
										c
010552	011494	012431	013364	014292	015214	016129	017036	017935	*018824	98
020234	022006	023761	025498	027215	028908	030576	032215	033821	*035389	96
029088	031584	034044	036463	038838	041164	043433	045641	047778	*049835	94
037152	040272	043330	046318	049230	052057	054789	057415	059921	*062295	92
044466	048116	051671	055122	058457	061664	064729	067635	070365	*072900	90
051066	055157	059116	062930	066583	070059	073338	076399	079220	*081777	88
056990	061439	065715	069798	073671	077312	080696	083799	086592	*089048	86
062273	067003	071513	075781	079783	083494	086885	089926	092585	*094833	84
066950	071889	076557	080929	084978	088673	091981	094868	097300	*099246	82
071056	076135	080891	085295	089314	092914	096059	098710	100834	*102400	80
074623	079781	084560	088928	092848	096282	099191	101536	103281	*104401	78
077684	082862	087605	091876	095634	098839	101450	103425	104733	*105354	76
080270	085415	090067	094185	097726	100645	102902	104455	105277	*105358	74
082411	087476	091987	095901	099174	101759	103613	104700	*104998	104509	72
084138	089077	093402	097069	100029	102236	103648	*104233	103978	102900	70
085478	090251	094351	097729	100338	102131	103068	*103123	102294	100618	68
086460	091031	094896	097925	100149	101497	*101932	101437	100022	097749	66
087110	091447	094992	097694	099506	*100384	100298	099239	097235	094372	64
087455	091528	094752	097076	098452	*098841	098219	096590	094000	090565	62
087518	091304	094184	096107	*097029	096914	095749	093550	090385	086400	60
087326	090802	093317	094823	*095276	094650	092938	090174	086450	081947	58
086900	090048	092182	*093257	093233	092089	089833	086516	082256	077271	56
086263	089068	090809	*091442	090935	089274	086481	082626	077858	072433	54
085437	087886	089224	*089410	088418	086244	082925	078554	073310	067492	52
084442	086526	*087454	087190	085714	083036	079207	074344	068662	062500	50
083298	085010	*085525	084810	082857	079684	075364	070040	063959	057508	48
082024	083360	*083460	082298	079874	076223	071435	065681	059247	052561	46
080637	*081595	081282	079679	076796	072684	067453	061306	054564	047703	44
079154	*079735	079012	076977	073649	069095	063451	056948	049948	042971	42
077592	*077797	076672	074214	070457	065486	059459	052641	045433	038400	40
*075965	075800	074280	071413	067245	061881	055504	048413	041049	034021	38
*074289	073759	071854	068592	064034	058304	051613	044292	036825	029860	36
*072576	071690	069411	065770	060844	054777	047808	040302	032785	025941	34
*070839	069605	066966	062965	057694	051319	044110	036465	028950	022282	32
069089	067518	064534	060191	054600	047950	040538	032800	025338	018900	30
067337	065441	062128	057464	051578	044684	037108	029322	021965	015805	28
065593	063385	059759	054795	048642	041537	033836	026046	018840	013006	26
063865	061360	057438	052197	045803	038519	030732	022982	015975	010506	24
062163	059373	055174	049678	043072	035642	027806	020138	013373	008305	22
060492	057434	052976	047249	040457	032914	025067	017522	011037	006400	20
058859	055548	050851	044915	037966	030341	022519	015134	008967	004782	18
057269	053721	048805	042682	035603	027928	020165	012977	007157	003441	16
055727	051958	046841	040556	033372	025678	018006	011046	005601	002360	14
054237	050262	044963	038538	031276	023590	016040	009338	004289	001521	12
052799	048635	043174	036630	029314	021664	014264	007845	003206	000900	10
046447	041558	035541	028694	021429	014286	007940	003147	000536	000000	00
1,22	1,50	1,86	2,33	3,00	4,00	5,67	9,00	19,00	∞	F/M
1,20	1,50	1,80	2,30	3,00	4,00	5,70	9,00	19,00	∞	F/K

Index